CULTURE CLASH

CULTURE
CLASH

Environmental Politics
in New Mexico
Forest Communities

A Memoir, 1970–2000

Kay Matthews

SUNSTONE
PRESS

SANTA FE

Sunstone books may be purchased for educational, business, or sales promotional use.
For information please write: Special Markets Department, Sunstone Press,
P.O. Box 2321, Santa Fe, New Mexico 87504-2321.

Book and cover design › Vicki Ahl
Body typeface › Garamond Pro
Printed on acid-free paper
∞
eBook 978-1-61139-291-3

Library of Congress Cataloging-in-Publication Data

Matthews, Kay.
  Culture clash : environmental politics in New Mexico forest communities : a memoir, 1970-
2000 / by Kay Matthews.
       pages cm
  Includes bibliographical references.
  ISBN 978-1-63293-005-7 (softcover : alkaline paper)
  1. Matthews, Kay. 2. Las Placitas (N.M.)--Biography. 3. El Valle Escondido (N.M.)--Biog-
raphy. 4. Community life--New Mexico--Las Placitas Region--History--20th century. 5. Las
Placitas Region (N.M.)--Rural conditions. 6. Gentrification--Environmental aspects--New
Mexico--Las Placitas Region--History--20th century. 7. Real estate development--Environ-
mental aspects--New Mexico--Las Placitas Region--History--20th century. 8. Social conflict-
-New Mexico--Las Placitas Region--History--20th century. 9. Environmental policy--New
Mexico--Las Placitas Region--History--20th century. 10. Las Placitas Region (N.M.)--Envi-
ronmental conditions. I. Title.
  F804.L28M38 2014
  978.9'57--dc23
                                  2014018823

WWW.SUNSTONEPRESS.COM
SUNSTONE PRESS / POST OFFICE BOX 2321 / SANTA FE, NM 87504-2321 /USA
(505) 988-4418 / ORDERS ONLY (800) 243-5644 / FAX (505) 988-1025

In memory of Mark Schiller

Contents

"If I can't dance I don't want to be part of your revolution."
—Emma Goldman

"There's a lot more to do in life than just writing."
—Grace Paley

Preface

This book was written over a very long period of time, sometimes as the events unfolded and sometimes later on, after the fact and with more reflection. Now, as it finally goes to press and I read and edit once again, the time frame expands even more and the past, present, and future blend together, reflected in both form and content: tenses get mixed up as I live, look back, and project while assessments change as I do.

I feel very fortunate to have lived in Placitas and El Valle when I did, essentially all my adult life. Both my children were born and raised there. The twenty plus years in each community have been extraordinary, signified by an array of people I wouldn't have met anywhere else, by events that challenged us in painful yet invigorating ways, and by personal growth and the pure pleasure of life in these villages for which I will always be grateful.

The changes, too, have been painful. The Placitas we left in 1991 was transformed from a small land grant community to a gentrified satellite of Albuquerque. El Valle has become a village of *ancianos* whose children have moved away (I'm included in that description); many of the men and women who were in their fifties and sixties when my partner Mark Schiller and I came, the life force of the village, are gone. Tomás Montoya, our closest *vecino* and village *mayordomo*, died in 2009 after being ill for several years with diabetes and cancer.

Then, in 2010, Mark died of pancreatic cancer. He lived for a year and a half after the diagnosis and in that time managed to finish a scholarly paper on the tenure of New Mexico Surveyor General George W. Julian and publish a small book of his poems. It's been a rough road without him, both emotionally and physically, as I work to maintain my independence and keep things going on my ten-acre El Valle home, where I still live.

We were together for 34 years, and most of what I relate in this book happened not just to me but to us. It is our story, and for this I am also grateful. But I must also note that while our story was inextricably linked to all those who lived and worked with us, I do not, as someone who came to these communities as an outsider, claim to speak for them. I hope this book helps give voice to the values and experiences we shared, but I can only tell the story from my perspective, not theirs. *Mil gracias* to all of you.

I also extend a special thanks to my sisters, Claudia and Riki, for their support of *La Jicarita* and this book.

"As in every kind of radicalism the moment comes when any critique of the present must choose its bearings, between past and future. And if the past is chosen, as now so often and so deeply, we must push the argument through to the roots that are being defended; push attention, human attention, back to the natural economy, the moral economy, the organic society, from which the critical values are drawn."
—Raymond Williams

"Pessimism of the intellect, optimism of the will."
—Antonio Gramsci

Introduction

New Mexico culture has been bought and sold many times. When I arrived in the early 1970s what passed as indigenous was once again "chic." Stores on Fifth Avenue in Manhattan sold turquoise necklaces, velvet shirts, cowboy boots, and Concho belts. The howling coyote, once a symbol of a wild and wily spirit in Native art, began to adorn T-shirts and baseball caps. Transplants from New York and California built million dollar "solar adobe" houses (often neither solar nor adobe, but a façade) perched on former empty hillsides where the coyote and jackrabbit ran free. Second homeowners in Santa Fe and Taos helped bolster a tourist based economy that resulted in both inflated land prices (which squeezed out the locals) and a cultural morass that was a parody of itself. It's a scenario that's been played out in the national arena since before New Mexican statehood: romanticizing and promoting "Spanish" and "Native" culture in the name of the very tourism and migration that ultimately devour local autonomy.

Although many battles have already been lost, and sometimes the future looks uninhabitable, Native American, Hispano, and Anglo communities work hard to remain intact under this onslaught of globalized consumerism. The impact of this economic hegemony upon already marginalized populations is revealed in New Mexico statistics: one of the highest rates of poverty in all the 50 states; increasing disparity between rich and poor; continuously high rates of teen pregnancy; and communities with the highest per capita heroin addiction in the country.

Yet in the villages of northern New Mexico, tucked away in the far corners down south, and in Santa Fe and Taos there are those who continue to pursue cultural and economic diversity and a "sustainable" life. What constitutes that sustainability—a word that is much abused these days by environmentalists, economists, and politicians trying to establish a platform—is a future that is more economically and environmentally just, largely what this book is about. It's about people who continue to grow food, be it a kitchen garden or a commercial crop—without pesticides. It's about maintaining the *acequias* in order to water that food. It's about knowing where to cut your firewood or how to get a thinning contract—and fighting to ensure access to the timber. It's about working at a job that provides a real service or fills a need, as opposed to one that manipulates money. New Mexico is one of the few places left in this country where people can integrate self-sufficiency and wage labor, subsidizing part time or low-wage jobs with the subsistence skills—raising cattle, cutting firewood, growing fruit and vegetables—they learned from ancestors rooted to the land for 300 years. These activities provide an alternative to the alienated labor many are forced to engage in to pay the bills and help people live consciously, responsibly, and humanely as a changing world economy creates new pressures to be technological, competitive, and increasingly work oriented.

For twenty years I lived in Placitas, the 1970s to the 1990s, a land grant community at the north end of the Sandia Mountains near Albuquerque. There, a complex alliance of Hispano land grant heirs, Sandia Pueblo, and back-to-the land Anglos tried to forestall the gentrification of the 1980s and 90s development boom, without success. In 1992 my family and I moved to El Valle, another land grant community further north in the Sangre de Cristo Mountains, where another alliance of Hispanos, Picuris Pueblo, and Anglos fought a weird mix of corporate and environmental gentrification in the forests and watersheds of *el norte*.

In this book I juxtapose chapters documenting daily life in these communities, the joys and difficulties, with chapters trying to figure out what exactly is at stake. Many modern and postmodern writers and philosophers have struggled theoretically with how society can integrate economic, social, and personal freedom and equality. *Norteños* are literally out there in the fields living the struggle. As society sinks deeper into cultural confusion I think it's important to document this struggle that transforms value, identity, and ideology into political action: "Indeed, much of the color and ferment of social movements, of street life and culture, as well as of artistic and other cultural practices, derives precisely from the infinitely varied texture of oppositions to the materializations of money, space and time under conditions of capitalist hegemony." (David Harvey, *The Condition of Postmodernity*)

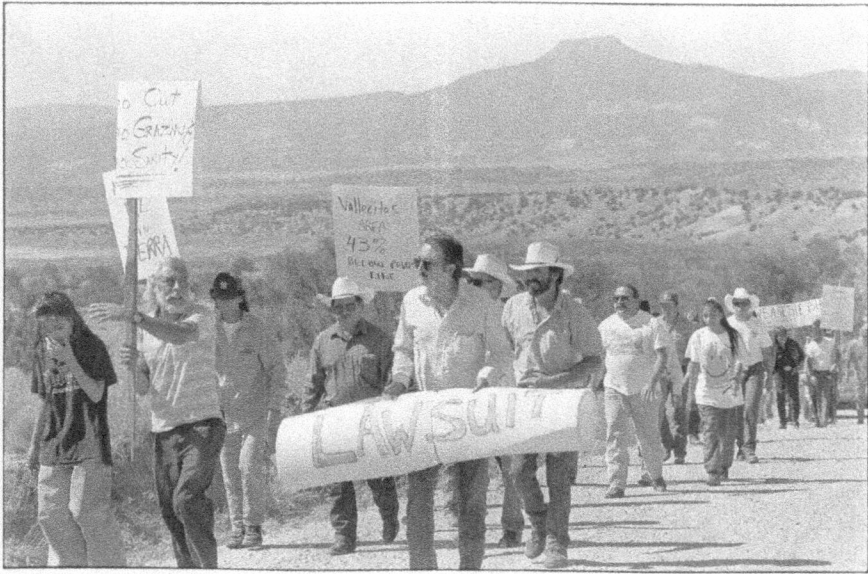

La gente marches on. Leading the march on the left, Ike DeVargas; on the right, Moises Morales. Photograph by Eric Shultz.

While some may think we are all hopelessly naive, railing against change that is inevitable, romanticizing and mourning the loss of a past that was perceived by the previous generation as its own loss, or was in its own right unjust and oppressive, a quote from the Robertson Davies book, *Rebel Angels,* sheds some light on our perspective: "The recognition of oneself as a part of nature, and reliance on natural things, are disappearing for hundreds of millions of people who do not know that anything is being lost. I am not digging into such things because I think the old ways are necessarily better than the new ways, but I think there may be some of the old ways that we would be wise to look into before all knowledge of them disappears from the earth—the knowledge, and the kind of thinking that lay behind it."

It's not only our relationship with nature that we need to understand and cherish. What we've forged relationally, in partnerships and friendships, and what we've learned together, is the measure of our success. It is this "continuous engagement" that gives our struggles relevance.

Origin

"A city was a great achievement, bridges were fine things to build. But the streets, harbors, spikes of stone were ultimately lost in the wider cradle of mountain and sky."

—Leonard Cohen, *The Favorite Game*

I first came to New Mexico from a home geographically near, yet culturally light years away. Colorado Springs, Colorado, where I lived for almost 18 years, is a suburban American community inundated with a military presence that keeps the government Republican and the atmosphere repressed. The town also became a mecca for the religious organizations of the moral majority, attracted by the conservative nature of the community and, I would venture to guess, a sympathetic town government willing to make their stay as pleasant as possible. Colorado Springs was also the home of Will Perkins, one of the movers and shakers behind the anti-gay amendment that passed the Colorado legislature many years ago. My mother once told me she fantasized about going to Perkins' car dealership and shooting him: she doubted they'd put an eighty-year-old in jail, but I think they probably would have.

My family took frequent trips to Santa Fe and Albuquerque when I was a kid. I always loved seeing the Indians who sat on the plazas selling their jewelry and pottery while Chicano lowriders drove souped-up cars up and down the streets, whistling and jeering to let everyone know this was their turf. Houses were not brick, two-story ranch style but made of earth, called adobe, that melded them with the land. The La Fonda Hotel, where in those days even a lower middle-class family like ours could afford to stay, had reddish-brown tiles on the floor, blue tiles on the walls, flowered tiles in the bathroom—and not a shag carpet in sight. Even the landscape was surreal: brown hills dotted with green, rising to red mesas of sandstone spires and pinnacles, giving way to high mountain peaks of lustrous blue and purple patched by lingering snows. The scenery in my native Colorado may be more spectacular, with its 54 peaks over 14,000 feet, but New Mexico is such a juxtaposition of diverse people and terrain that it seduced me immediately, like it has so many others before me.

It would be the first place I settled after I was on my own. Before that, though, I had to attend a college as far away from home as possible, with as much prestige as possible, and the most progressive curriculum I could find to compensate for four years of torture at Roy J. Wasson High School. I enrolled at Antioch College in Yellow Springs, Ohio. Talk about wild, heady, and free. Antioch was all of those things and more, in its mid-America setting of rich black farm fields, hardwood forests, and stately brick buildings housing a faculty that at one time included jazz great Cecil Taylor. It was all there for the taking—coed dorms, no grades, interdisciplinary study, co-op jobs all over the country, endless opportunities for creative learning and growth.

Unfortunately, I, like many others there, essentially "freaked." Eighteen year-olds from provincial backgrounds aren't quite capable of making many mature decisions when faced with academic, social, and political freedom. Even the supposedly more sophisticated New York kids (I was immediately intimidated by their assertive manner and worldly demeanor) who came from Hunter High School and the School of Performing Arts didn't necessarily deal well with Antioch's lack of structure. Maybe by the time students were 4th or 5th year (Antioch lasts five years to accommodate co-op jobs) you were able to settle down and capitalize on some truly outstanding academic opportunities, but many of us didn't make it beyond the sex, drugs, and rock 'n' roll. The attrition rate was almost fifty per cent.

On top of all this the year was 1968 and the country was erupting in political madness in Chicago, Washington, DC, San Francisco, New York, and our own little Yellow Springs campus in central Ohio. While I was enrolled at Antioch I participated in two Vietnam antiwar demonstrations in Washington (the largest ever in 1969), numerous local demonstrations in Cincinnati and Columbus, and was tear-gassed in Berkeley on a co-op job. The entire Yellow Springs campus shut down to protest the bombing of Cambodia, the killings at Kent State, and a local strike of campus workers (we tried to be democratic in our political affiliations). The Afro-American dorm closed its doors to white people, including some southern whites on Rockefeller scholarships who proceeded to try and shoot their way in, and the SDS'ers and YSA'ers fought constantly over whose ideology was the more "politically correct."

The two years I spent at Antioch were filled with political, sexual, and drug-taking escapades at the expense of any academic achievement. But that was all right, really, in the scheme of things. The $3,000 a year tuition my parents paid (half of it was scholarship and loan) was well spent, I think, on a social and political education that took its toll on my sanity while there but in the long run has proved invaluable. Antioch was a transformative experience for me and many like me.

In those days we hitchhiked everywhere. And wherever we ended up, there always seemed to be someone to take us in—a fellow student on a co-op job, strangers in a house where someone on a co-op job *used* to live, students from other universities who themselves had been out hitchhiking and were taken in. As a safety net for its students, Antioch published what they called a moocher's guide, the names of students' parents who didn't mind other students dropping in to spend the night when they found themselves stuck in Topeka, Kansas or Peterborough, New Hampshire. It was best if you pretended to know the student whose house you were staying in, to have something to talk about with the parents, but none of us ever hesitated to use anyone in the moocher's guide as last resort.

In Berkeley I hitchhiked back and forth to work at an "experimental" private school from my house on Tenth Street, by myself, until the guy in the Oldsmobile asked if he could make me happy by showing me his erection; from then on I commuted by thumb with one of my housemates, a teacher's aid at Berkeley High. I hitchhiked back and forth across the country with various friends, getting rides from those folks whose lives were much more precarious than ours (without middle class parents to bail them out)— hippies, truckers, Okies, the homeless. One time a friend and I got a ride on an eighteen-wheeler hauling timber from Mendocino to Los Angeles. Instead of turning off in San Rafael to sideswipe San Francisco, our taciturn driver took us across the Golden Gate Bridge into San Francisco and down to the Mission District where our friends lived in ethnic squalor. "Wouldn't want my daughter out there on the highway without someone like me to pick her up and deliver her safe and sound," he said. He wanted to take us right to the door but he couldn't manage the turn onto 19th Street from Market.

Occasionally our parents would loan us their cars if they knew it would keep us off the streets. One winter my best friend and I borrowed her parents' car to take boyfriends and their two motorcycles—on a trailer behind the car—over the Rocky Mountains to their co-op jobs in Berkeley. We drove straight through one blizzard after another, stopping Christmas Eve at a diner in Reno, over Donner Pass before they closed the road, down to the flower-lined highways of Sacramento into seventy degree Berkeley.

With that same friend and another male friend I once borrowed my mother's car to take a trip to New Mexico. Driving into Taos over a high mountain pass we ran into another group of Antiochians pushing their car *out* of Taos after it died at 9,000 feet. Later, driving out of Albuquerque toward Santa Fe we picked up a hitchhiker who told us he was on his way to some commune in Placitas and that we were welcome to spend the night there. We drove him up into a small village in the mountains, left the car on the highway, and walked across the creek to a crumbling adobe where we spent the night on the floor along with ten other people. Several years later I found out we had spent the night in the house of Ulysses S. Grant, who later shot several hippies over a dope deal, rode out of town on his white horse, and hasn't been seen or heard of since. I would subsequently live in Placitas for twenty years.

Arrival

> "The 'beat' movement created the most important breach in the solid, middle-class values of the 1950's, a breach that was widened enormously by the illegalities of pacifists, civil rights workers, draft resisters and longhairs. Moreover, the merely reactive response of rebellious American youth has produced invaluable forms of liberation and utopian affirmation—the right to make love without restriction, the goal of community, the disavowal of money and commodities, the belief in mutual aid, and a new respect for spontaneity...."
>
> —Murray Bookchin

I dropped out of Antioch in the spring of 1971 and got a job working for the Forest Service in Cloudcroft, New Mexico. This was during the days when government agencies were scrambling to fill quotas, and I was hired to be the token woman in a male bastion of beer drinking, pool-playing cowboys who lived to fight fire. My job was to patrol the Forest Service campgrounds, collect fees, and do general PR work—it was years before women were hired to be fire fighters, which was all right with me, as I saw right away that it was hot, dirty, hazardous work. Unlike a lot of people I know, my adrenalin doesn't have to pump a mile a minute for me to feel like I'm alive.

Cloudcroft is in the southeastern part of the state, which belongs to Texas, really, along with much of the rest of southern New Mexico (El Paso, Texas, the border town, really belongs to New Mexico). In the summertime the campgrounds around Cloudcroft fill up with Texans in their campers, retired folks moving from campground to campground every two weeks (the time limit on each site) to spend the summer in the cool mountains. The locals are mostly Anglo ranchers, except for the Mescalero Apaches, whose reservation borders the national forest. They're ranchers, too, but are imbued with the memories of ancestors who first lived on these lands when open range and huge herds of deer gave them freedom and renewal.

I fit in easily enough. When I showed up in my two-cycle Saab and calf-length skirts I'm sure they were worried, but I did my job and managed to fall in love with one of the fire fighting cowboys who taught me how to play pool, ride a horse, two-step, and ferret out commonalties in someone so different from me it was embarrassing when I thought about it too much.

Cloudcroft did little to prepare me for Santa Fe, though. When my Forest Service job was over I retreated home, to Colorado Springs, where I hung around trying to think of what to do next. I was spared too much worry when a friend from Antioch showed up and told me he had just set up a co-op job with the Sierra Club in Santa Fe and I should apply before he reported it to the college and the other students also had a chance to apply. I got the job and two weeks later was headed south on the bus, through the small Hispano towns of southern Colorado (which ideally should be part of New Mexico). Brant Calkin picked me up at the bus station and took me to his hilltop home in Pojoaque. Brant was probably the only person in Santa Fe who in 1971 still wore his hair in a crew cut. The last time I saw him he was running for State Land Commissioner on an environmental slate, and after losing he moved to Utah to head up the Southern Utah Wilderness Alliance out of Salt Lake City.

Back then, as New Mexico head of the Sierra Club, Brant liked to project the image of renegade, although the crew cut definitely confused the issue. While the Sierra Club could never be considered a renegade outfit, even by the government agencies it constantly did battle with, Brant *was* supportive of a lot of renegade groups that worked under the umbrella of the Central Clearing House, housed in the same office as the Sierra Club. One group was off fighting the strip-mining on Black Mesa, sacred mountain of the Hopi Indians, while another group fought the proliferation of subdivisions sprouting up around Santa Fe, a trend that would eventually doom much of the state to what I considered cultural and aesthetic death (and which I would spend years battling in Placitas). The two hippies who founded The Famous Barter Brothers Co-op, a friendly place offering healthy food at reasonable prices without exploitation of workers or clientele, used our office as business headquarters whenever necessary. And the whole operation was underwritten by Harvey Mudd, whose guilty conscience or sense of fair play—he is the descendent of a copper mining magnate—benefit us all. Harvey lived in Arroyo Hondo (outside Taos) in those days, with a Mexican wife, and came down to Santa Fe occasionally in his Porsche to survey the situation and mingle with the workers. At some point he moved to San Francisco to write poetry.

The array of interesting and weird people was endless, the atmosphere was charged with cooperation and good will, and Claude's was still there. That was Santa Fe's hippie bar, similar to every other hippie bar in every other port in America that will be remembered by all of us who came of age in them. One of the best—and one of the few that remain—was in Madrid, out in the foothills southeast of Santa Fe. In these early '70s, the hippies had already started to move into this old run-down mining town, full of weathered clapboard houses and obsolete mine shafts. A few folks sold various wares out of their patched-together houses and congregated at the Mine Shaft Tavern for fun and games.

While I was at the Central Clearing House we held a fundraiser, in conjunction with the Sierra Club and Zero Population Growth, out at the Mine Shaft Tavern. Sally Rogers was the mover and shaker behind Zero Population Growth, full of energy and zeal. Sally continued to work in mainstream environmental movements in Santa Fe, and eventually became a high-profile figure, along with her New Mexico Representative husband, Max Coll, in New Mexico politics. Lisa Gilkyson (now known as Austin's own Eliza Gilkyson) and her band, The Family Lotus, was the featured music at the benefit. I can't really recall how much money was raised, but I know we all had a good time dancing in the beautifully restored bar, fraternizing with the Santa Fe elite.

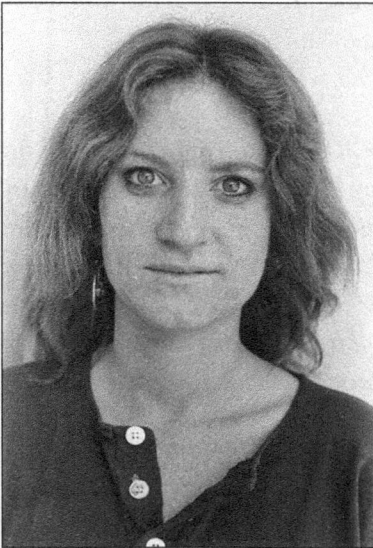

Alternative newspaper Seer's Catalog staff photograph of the author at age 23. Photograph by Tony Louderbough.

I stayed on in Santa Fe after my co-op job ended, due to Brant's generosity, but eventually strayed on down to Albuquerque, sixty miles south, with some notion of once again pursuing an academic career. I did take some English and Women's Studies classes at the University of New Mexico, but mostly I wrote for several alternative newspapers and hung out with a bunch of people who were fighting all the subdivisions around Albuquerque. AMREP Corporation, out of New York, was developing Rio Rancho, then a few houses surrounded by miles of bladed desert roads, drowning in dust. AMREP marketed the whole operation as a land investment scheme, bringing in people from back east to buy lots in the subdivision as an investment, which they would be able to sell at profit in just a few years as more and more people moved to this great new community in the middle of the desert. What AMREP neglected to tell these investors was that the company owned so much land itself—thousands and thousands of acres—that the supply would always outlast the demand and there would be no profit. A couple of AMREP executives eventually went to jail for land fraud (former governor Toney Anaya eventually pardoned them), but Rio Rancho survived and grew beyond our wildest expectations. All the people AMREP brought out to invest in worthless sand decided instead that they wanted to *live* on the sand, and today Rio Rancho is an incorporated city of many thousands of people. It is perhaps the most ill conceived, aesthetically (and water) challenged, and regressive city in all of New Mexico, but it is indeed there.

Dave Foreman

Dave Foreman was one of those Albuquerqueans fighting AMREP, after his stint as a young Republican for Nixon and before his tenure as Wilderness Society lobbyist in Washington, DC. Now that he's credited as the founder of Earth First! and touted as the radical of the environmental movement, the former affiliation is somehow lost in the biographical data that explains his radicalization as a reaction to his frustrations as a mainstream lobbyist. I think it's impossible to ignore the Republican stint. As a matter of fact, Foreman used to go around saying, "You can be a Republican and still be an environmentalist."

Sorry, Dave, but you can't. Neither can you be a redneck, good old boy, or purveyor of the Earth First! macho image (however tongue in cheek it may be at times) that encourages the kind of diatribe Foreman mentor Edward

Abbey leveled at "illegal aliens" as diluters of American culture, or Earth First!'s Malthusian philosophy that claimed AIDS is a natural population control. Apparently some other Earth Firsters began to realize that human exploiters are also the exploiters of our planet—the corporate imperialists who depend not only upon the cheap labor of the world to be their grunts but the natural resources to supply the material. Foreman subsequently quit Earth First! because it moved "too far left."

While Albuquerque had its own hippie bar, Okie Joe's, we used to drive up to the Thunderbird Bar in Placitas, a small Hispano village about twenty-five miles away at the north end of the Sandia Mountains, for the quintessential hippie bar experience. A rambling adobe structure with huge *vigas* and windows, the Thunderbird had it all—a tap run by locals, a community kitchen with weekend guest chefs, a jukebox, TV, pool table room, and live music by Tracy Nelson and Mother Earth, Sonny Terry and Brownie McGee, and New Mexico's own Cadillac Bob. The Thunderbird was infamous, attracting people from not only all over New Mexico but the country as well. Much of this was due to the communal scene that existed alongside the good-time scene. Placitas was the site of the solar dome commune Drop City South, as well as Towapa, Sun Farm, and Lower Farm (where Ulysses S. Grant lived). By the time I arrived, in 1973, the communal owners of Drop City South had defaulted on the property, and the other communes' original members had left or moved on to private dwellings near their former communal digs.

I moved into a large adobe house in the village with a friend I had met on the battle lines in Albuquerque. Surrounded by five acres of orchard, the Sandia Mountains out my back door, and the Thunderbird Bar a short walk down the road, I thought I had found heaven. We all did, I guess, back in those days of cheap rent (mine was $50 a month), good music, and endless possibility. While many others sought their personal and social salvation at New Buffalo in Arroyo Hondo, the Hog Farm in Llano, crumbling adobes in Truchas, Mora, and Guadalupita, I followed suit in Placitas.

Las Placitas

"Most of the luxuries, and many of the so-called comforts of life, are not only not dispensable, but positive hindrances to the elevation of mankind."

—Henry David Thoreau

Humans have occupied Las Huertas Valley for thousands of years. In the lower valley, where the present day village of Placitas lies, Pueblo Indians from villages in the adjacent Rio Grande Valley built small seasonal farming settlements. At the time of Coronado's arrival at the banks of the Rio Grande in 1540 there were no permanent Indian settlements in the Placitas area.

While archaeological remains indicate that several Spanish colonists lived in the area in the late 16th and 17th centuries, it wasn't until the mid-18th century that Juan Gutierrez petitioned the Spanish governor of New Mexico for a tract of land there for himself and eight tenant families. Two years later, when the petition was granted, 21 families were living in Las Huertas, about a mile northeast of where the village of Placitas is now.

Lynn Montgomery

Lynn Montgomery lives in the remnants of the Sun Farm commune, adjacent to the original Las Huertas community on lower Las Huertas Creek. Until several years ago, no road accessed Lynn's property across the creek, and there is still no electricity for lights and water. Lynn irrigates from a spring that feeds the creek and runs his computer with solar cells. Former president of Sandoval County Environmental Action Community and a dedicated farmer, Lynn has struggled long and hard to preserve the Las Huertas site. Twenty-two acres are now owned by the Archeological Conservancy, but the remaining 10 acres of land were for many years in ownership dispute between Placitas people who lived there—communal drop-outs, down -and-outers, families with little money and no other place to live—and some Albuquerque lawyers who wanted to subdivide the land. Lynn wanted them all gone.

For the original settlers, life on lower Las Huertas Creek was not easy. In 1810, with the outbreak of the Mexican War of Independence, Spanish forces were diverted from the northern frontier, and the Apache and Navajo began raiding the settlement. In 1823 the residents of Las Huertas were ordered to abandon the village by the Spanish governor, unable to provide them protection from the intensification of the raids. Many of the families moved to Algodones, just north of Bernalillo along the Rio Grande. Ten years later the families started moving back to the area at the north end of the Sandia Mountains, settling in four communities, with the present day Placitas as the population center.

The end of the nineteenth and beginning of the twentieth centuries were also difficult times for the people of Placitas, as they were for the other land grant communities all over the state. The creation of the United States Court of Private Land Claims, ostensibly created to sort out common and individual claims to unsurveyed lands, required that land grants defend their land titles. San Antonio de Las Huertas did as so many land grant communities did—they hired attorneys who were familiar with the intricacies of this new law (and who could speak English) to defend their title in court. What ensued constitutes one of the ugliest chapters in New Mexico's history, and the names involved read like a who's who in the territory's politics—Catron, Bond, McMillen, Otero. In payment for services that many of the land grant heirs never understood, and for services often rendered in collusion with the court, the grants relinquished their lands. In a culture largely based on barter, with little if any capital to spare, land was the only negotiable payment. San Antonio de Las Huertas Land Grant lost the eastern third of its grant to the lawyers.

While the loss of land to nefarious lawyers was bad enough, many of these former communal grant lands eventually ended up in the hands of the U.S. government. By the beginning of the twentieth century, the results of America's rush to conquer the West was all too visible. Horace Greeley's exhortation, "Go West young man," had been heeded by miners, trappers, loggers, and ranchers who came into these lands of plenty in unprecedented numbers, hell bent on harvesting the vast resources as quickly and profitably as possible. The beaver was nearly extirpated from the mountains in a matter of years. Hillsides were stripped of timber in a matter of days, causing massive soil loss and silt-filled streams. Big time ranchers came and unleashed thousands of head of cattle, fencing out sheep while their cattle overgrazed lands and degraded watersheds.

The land grant people, along with the Pueblo Indians who lived in or nearby the mountains, had long used the forests for sustenance: gathering fuelwood, hunting the prolific populations of deer, elk, bear, and mountain lion, grazing their sheep. While some of the grazing and timbering practices employed by the local villagers caused erosion and pollution problems, most of this abuse occurred on or near their private property in the villages themselves, and was on a relatively small scale. To counteract the much more pervasive devastation wrought by the unregulated activities of the loggers and miners, the federal government, under President Teddy Roosevelt, created the Forest Reserves. With the passage of the original National Forest Management Act and Roosevelt's transfer of the reserves to the Department of Agriculture's Division of Forestry, the Reserves became the Forest Service. San Antonio de Las Huertas Land Grant essentially lost the north end of the Sandia Mountains, which became part of Cibola National Forest in 1908. As farming and ranching became even less profitable, many villagers left Placitas to seek work in Albuquerque, twenty-five miles away. This proximity allowed residents to earn wages without having to leave home for months at a time, as residents in more isolated northern New Mexico villages often had to do, and also provided better educational opportunities.

The village remained a Hispano enclave until the 1940s, when Albuquerque began to grow. "I think the first Anglo family moved up here around 1948," Lizzie Archibecque told me. Lizzie ran the grocery store in the village after her mother, Doña Aurelia Gurule, retired to tend her gardens and raise goats, geese, ducks, and horses. "Some of the people who moved into the village opened little stores, and Edna McKinnon ran a really nice restaurant. I think she called it La Casa de Las Huertas. I started working for her as a waitress, cook, dishwasher, everything, when I was 16."[1]

Lizzie Archibecque

Lizzie's mother died in 1991. Lizzie stills grieves. She and her husband Jake kept the store open for a few more years, but were always threatening to close. It seemed like Lizzie had always been there. If you wanted to find out what the siren was you heard the night before, or what happened at the community meeting with the Forest Service at the school, all you had to do was go to Lizzie's. She also was a soft touch, often letting people run a tab at the store when they were short of cash. We all knew that she and Jake were worn out and feeling their store days were numbered. Even though most of the people who lived down in the snazzy new subdivisions didn't frequent their store, Lizzie and Jake knew that when the strip shopping mall went in alongside the highway, things would be different. Then, in the fall of 1992, someone built a mini-mart across the road from Jake and Lizzie's. Finally, in the summer of 1996, they closed. It was truly the end of an era.

One of the first of the Anglo immigrants, a Minnesotan of proportions as large as his home state, was John Nordstrom, a man who was to provide a link to what the '60s and '70s would bring to Placitas. John came to New Mexico in 1957 to study anthropology at the University of New Mexico but soon abandoned Albuquerque for Placitas, where, with another transplanted Minnesotan, Ralph Roller, he bought 30 acres from the Las Huertas Grant in a long, narrow valley south of the village.

"I guess I was one of the first subdividers up here," John said shortly before his death in 1986. "I bought the land for $500 or $600 an acre and split it up with the lay of the land in mind."[2]

The 1960s brought hundreds of counterculture advocates to the area. They started the communes that included Drop City South; boasting domes constructed of multicolored junked car tops, this commune was the prototype for the more famous Drop City near Trinidad, Colorado. "I sold the land to Drop City South," John said. "The commune lasted about two years, when the founders were left holding the bag and defaulted on the property. I lived in the triple dome while I built my adobe house down the road."

John was also one of the founding fathers of the Thunderbird Bar, along with Ralph Roller and Manny Nieto, another Placitas resident. He financed it, participated in the building, and was a regular guest chef at the bar.

John and Tom Nordstrom

John used to get me to bake apple pies for him in exchange for his famous bar chile. He stroked my ego by telling me the pie was the best he'd ever eaten. John had a heart as big as the rest of him. Once he had moved into his adobe house he rented the original solar dome to Mark, the man who would eventually be my partner, at a cheap price. Mark always said that if he ever ended up in jail, for whatever reason, John would be the first one he'd call to bail him out.

John also took care of his son Tom, Tom's wife Kimiko, and their kids, who lived in the triple dome. Tom had one of the most bizarre histories I ever heard. Separated from John at age seven by divorce, Tom grew up in Rhode Island with his mother and Jewish stepfather, Ed. At seventeen Tom ran away from home, crisscrossing the country, collecting adventures like emblems. Meanwhile, Ed divorced Tom's mother and emigrated to Israel, where Tom decided to join him. He was immediately drafted into the Israeli army and was stationed in the Sinai, protecting this newly acquired territory won in the Six Day War. Then John showed up in Israel and reconnected with Tom. When Tom was released from the army, he met Kimiko, from Japan, on an Israeli kibbutz, followed her back to Japan, married her, and brought her to Placitas, where John gave them the triple dome. Ten years later, Ed moved to Placitas and moved into the solar dome.

Unfortunately, John died prematurely, a victim of a lifelong addiction to cigarettes and alcohol. He took care of everyone but himself.

Many of the Hispanos in Placitas were uneasy with the new lifestyles and defensive of their own culture and traditions. Some were offended by drug use, dress style, and the idea of unmarried people living together. However, while shoot-outs and confrontations occurred in some northern New Mexico communities like Taos, Placitas saw few acts of violence between Anglos and Hispanos. "We didn't even know what hippies were," Lizzie said, "but we got used to them pretty fast."[3]

The Placitas residents may have been more tolerant of the newcomers because proximity to Albuquerque exposed them to a diversity of lifestyles, but it was also apparent that many of those who came to Placitas were attracted to the life and values of the Hispanos. I think the distinction can be made between these newcomers and the "hippies"—flower-children, druggies, whatever—who came with the naive notion that by taking drugs and wearing long hair and minimal clothing peace, love, and harmony would envelop the earth. The "back-to-the-land" or "alternative generation" folks were a little more sophisticated and politicized as to what it was they were trying to achieve by moving to Placitas or any of the small rural areas in New Mexico to which they came. I would have described my own motivation for being there—naïve in its own way, but heartfelt—as a rejection of bourgeois consumerism. The distinction is also a matter of class: many of the hippies who showed up in the Placitas environs came from hard scrabble backgrounds looking for a substitute family. The mostly middle-class back-to-the-land immigrants wanted to be connected to an environment, a home, a community, just as the local people were tied to the San Antonio Land Grant for hundreds of years. And because John Nordstrom and Ralph Roller offered land for sale at a reasonable price, people with little money but great notions about living a simple, healthy, environmentally sound life were able to move into Placitas. Their intent was not to change the nature of the Hispano community, but to integrate into a village they felt exemplified those values. Most of the hippies, who peopled some of the communes but never had the money to acquire land of their own, left the area, or became marginal members of the community. Unfortunately, the simple fact that the alternative generation *came* opened the door to the *next* invasion, which was not to be so benign.

The Forest Service—A Destructive Dominion

"From the earliest day in its history, the Forest Service has held the views that its obligations were not to the American nation, but only to that section of the public desiring the most unrestrained consumption of the national forests for their immediate financial and political advantage, without regard for the future needs of the nation."

—Willard G. Van Name, associate curator emeritus, American Museum of Natural History

When I was in Cloudcroft I once visited the woman who worked in the district fire lookout tower and was immediately captivated by the austere setting and romance of her job. Then my cowboy boyfriend took me up to visit an abandoned fire tower where we fantasized about living together, watching for fires and making love under the stars, and I was hooked. I vowed never again to patrol campgrounds or fight fire (women were slowly being allowed on the fire crews), and when I again applied to the Forest Service for work in 1975 I was lucky enough to be hired as the lookout on La Mosca Peak, near Grants, New Mexico.

La Mosca sits just below the ancient volcano, Mount Taylor, the tallest mountain in central New Mexico, 11,300 feet high. Called Tsoodzil by the Diné (Navajo), this mountain is at the center of their universe, where humans emerged to take their place in earthly society. I lived in the power zone of the mountain: above the alpine forest of spruce and fir, the aspen meadows full of browsing deer; below the flight patterns of a pair of golden eagles, riding the air currents in search of prey. In my lookout I was left alone, as long as I reported all my "smokes" and directed the crews toward the fires. That was not so easy sometimes, to figure out where exactly the fire was in a sea of green forest. But I usually managed to get the crews into the right canyon or onto the right road, and I loved my lookout, its solitude and repose.

But the image of the inaccessible fire lookout, ensconced in a tower of isolation to watch over the lonely woods, was only valid in the stories of Jack Kerouac and Gary Snyder. On Mount Taylor, across the canyon from my lookout, the exposed slopes of the mountain's 1950 clearcut presented a mess of fallen logs, brutalized stumps, impassible thickets of gooseberry and raspberry bushes, and washed out gullies. Underneath my mountain lay miles of tunnel accessing one of the country's largest veins of uranium ore waiting to be turned into yellow cakes of oxide to fuel nuclear power plants. Everywhere I looked roads carried ten-ton rigs of drilling equipment busily proving underground claims of their own.

Drilling rig on Mount Taylor.

Originally called forest reserves, the Forest Service was established by the federal government to protect large areas of western lands from the unabated mining, timbering, and grazing abuses of the late 19th century. Various interests debated the kind of management the reserves should see—preservation as aesthetic resources or preservation for commodity resource production—but until the passage of the original National Forest Management Act and Teddy Roosevelt's transfer of the reserves to the Department of Agriculture's Division of Forestry, the debate was largely academic. When Gifford Pinchot was appointed head of the reserves, and changed their name to national forests, the management philosophy advocating long-term timber and resource production prevailed.

Before World War II the main business of the Forest Service was essentially the same as when private interests held sway—timbering, grazing, and fire control—and lands under Forest Service management were extensively altered and abused. After the war, as skyrocketing timber demands devastated even more forest resources, the government was forced to take a closer look at the state of our public lands. The Multiple Use-Sustained Yield Act was passed in 1960, introducing the concept of multiple use for the first time in Forest Service management. This act required that *all* renewable forest resources be managed so their productivity not decline. The Forest Service image was deliberately altered to present a picture of preservation— conservator of wildlife and wilderness as well as timber and rangeland— rather than a purveyor of clearcuts, silt-filled streams, and eroded hillsides.

In reality, until quite recently multiple use did little to change the way the Forest Service did business because the multiple use that remained at the heart of Forest Service activity was timbering—its *raison d'etre*. In any forest, on any mountain, in any state, you can see the devastating results of Forest Service timbering policy that I could see from my lookout. Hundreds of years are required for a forest to recover from a clearcut, if there is a chance of recovery at all. Natural succession—the growth of grasses and bushes, next aspen and hardwoods, then pines and fir—will eventually repopulate an area with its original flora (and hopefully fauna), but the Forest Service, in an attempt to imitate nature, has replanted many of the clearcut areas with one kind of tree, precluding any kind of natural succession. The trees become susceptible to the disease and fire that a normal, diverse forest can tolerate.

Evenaged management, the most utilized method of timber harvest, perpetuated this process. In this type of management the same age trees within a stand were harvested by a rotation system of removal at 10 to 20 year intervals, utilizing both shelterwood (leaving some of the cone bearing trees standing) and clearcutting techniques.

The Forest Service not only defended its specific timber policy but promoted timbering in general as the foundation for all multiple use activity. Despite heavy subsidization and millions of dollars lost annually on timber sales, Forest Service officials claimed that timbering activity supported and made possible these other uses, justifying the timber losses. Ironically, one of the most egregious claims was that large-scale timber harvests were necessary to counteract the effects of 80 years of fire suppression, another Forest Service policy that, like that of the Park Service in Yellowstone, created a behemoth of potential destruction due to accumulated ground fuels that nature had previously taken care of with wildfires.

The Forest Service also claimed that without clearcuts and other extensive timbering operations there would no longer be the visually appealing and preferred wildlife habitat of the open meadow, once cleared by wildfire. Wildlife specialists pointed out that the various forest management plans (stipulated by the National Forest Management Act of 1976) calling for double and triple timber yields were *not* justified in wildlife terms. The increased use of cable logging on slopes of over 40 percent (using a winch system to haul the logs up the steep slopes) allowed harvests of older timber, some of which had never been previously cut. This left fewer snags for wildlife and reduced the habitat of other species, such as the black bear and spotted owl, which require a 70 percent crown cover. Wildlife specialists felt that as one of the multiple use resources of forest planning, their field was rarely given equal consideration. Because it's harder to put a monetary value on wildlife than timber or minerals, and the name of the budget game was to prove your worth, this particular multiple use had a harder time competing.

From a purely economic standpoint, the timber industry readily defended its Forest Service subsidization by listing the contributions the industry made to the public economy. It accused critics of failing to credit the industry's impact on local economies, through jobs, and county economies, through paybacks. It liked to act tough and threaten that any cutbacks in the industry would prove disastrous—despite the fact that 85 percent of the nation's lumber comes from private timber stands. Most of the 155 forest plans promulgated in the 1980s called for dramatic increases in timber production, and the Reagan and Bush administrations hoped to see a doubling of the national cut.

Most of the attention to these huge increases was focused on the timber heart of the country, the Pacific Northwest. There the battle lines were already drawn. The timber industry was charging that the environmentalists—seeking to reduce harvest levels and prevent the cutting of old growth forests—were the evil, narcissistic elitists taking the food out of the mouths of loggers and mill workers. Environmentalists were hiding behind the spotted owl when they should have been accusing the timber industry of obfuscating the *real* economic issue—that lumber mills were closing because of changing markets, obsolescence, and competition from cheaper milling in Japan. Loggers and mill workers were asking what they were supposed to do for a living once the mills closed and the lands were depleted. The *politicians,* who should have been the ones recruiting new kinds of clean industry to supplant a dying timber industry, and calling for an emphasis on smaller, community-based logging, were scratching their heads and blaming everyone who wasn't a possible constituent.

None of these problems would be solved until the Forest Service radically altered its management policy to reflect the changing nature of society and the fact that there would no longer be a guaranteed sustained yield of timber on public lands if timber harvest levels were not reduced. While the Multiple Use-Sustained Yield Act recognized the need to ensure the productivity of *all* forest resources, as long as the key word *productivity* remained at the core of Forest Service philosophy, industrial, commodity production would always take precedence over preservation and sustainability. The Forest Service had to quit giving in to special interests, which insisted that the timber industry could not remain viable without raping the land and rationalizing the results. It has only been within the last few years, due to national political pressure, largely in the Northwest, and constant pressure from grassroots environmental groups, that many forest districts finally began to change their *modus operandi.* Here in New Mexico, where so much of national forest land is former grant land, the Forest Service has been forced to address issues of sovereignty and access, which I will discuss in detail in later chapters.

Settling Down

"Never give holiday to the water."
— Jacobo Romero, in William deBuys' *River of Traps*

I n 1974 I rented Anne Rustebakke's Placitas house, an old added-onto
adobe that commanded a prestigious place in the village and a startling
view of the Jemez Mountains to the north. Anne, one of the older
Anglos who had been living in Placitas for many years, was running a hostel
up in Jemez Springs and needed someone in her house who was willing to
watch over her incredible New Mexico collection of pots, rugs, and artifacts.
I shared the house with Marsha, a native New Mexican, daughter of a
notorious local cattleman and sometime gambler who apparently made and
lost several fortunes during his lifetime. Marsha and I attended the University
of New Mexico in a rather haphazard fashion, taking mostly 20th century
literature classes to avoid Pope and Milton. Marsha eventually graduated; I
got sidetracked by the Placitas experience.

While most of the communes were defunct by the mid-1970s, Placitas
remained a thriving alternative community. At that time it consisted of the
village, where most of the San Antonio de Las Huertas Land Grant heirs
lived, and a few outlying communities such as Placitas West and Dome
Valley, where former commune dwellers lived with families and friends in
more traditional groupings. The village of Placitas is a green oasis in the
middle of the high desert landscape of piñon and juniper that distinguishes
so much of New Mexico. Watered by two spring-fed reservoirs in the Sandia
Mountain foothills above the village, cottonwoods, elms, fruit trees, and
gardens create a lush background for the array of dwellings that house the
human population: old adobes, trailers set up next to adobes, plastered-in
trailers made to look like adobes. Lizzie's store, the Thunderbird Bar, the
Catholic and Presbyterian churches, the elementary school, and post office
were the only nonresidential structures until the Laughing Bear Gallery and

the Placitas Land Company office appeared to signify the beginning of the end.

The Laughing Bear Gallery (a front for the factory where the owner turned juniper boughs into oil and sachet) sprung up across the arroyo from Larry and Lenore Goodell's house, causing untold resentment and bile from these two longtime Placitas residents who have become a legend in their own time. Larry, a performance poet who worked at the Living Batch Bookstore in Albuquerque, and Lenore, a painter and photographer who worked as a graphic designer, are almost synonymous with the Placitas alternative experience, having lived in the area for over forty years. They are the litmus test for the kind of people we hoped Placitas would always be home to—creative, community minded, interested in growing trees and vegetables, raspberries and flowers, good cooks, and famous party givers. Never mind that they are both a little crazy and irascible. Who wouldn't be, after what they've seen come down the line.

Larry sits at his piano pounding out a song:

Take me for a walk in Placitas
where have all the gardens gone
where have all the people gone to
I guess they're out on the run—
rushing from the town to the village
to get all the do re mi—you
can't have any fun without it
they say it's what makes their life free—

Take me for a walk in Placitas
where have all the non-rich gone
they've moved far away where it's cheaper—
workers, artists who lived here long
now there's just the big bulky houses
no one grows a thing—there's no time
and inside are xerox copy yuppies
pretending their life is sublime.[4]

Many other artists have called Placitas home. The poet, Bob Creeley, lived on and off in Placitas throughout the '60s and '70s in a rambling adobe always threatening to fall apart. His former wife, Bobbie Louise Hawkins, a writer, has toured with Rosalie Sorrels and Terry Garthwaite, a trio of inimitable women. Ed Dorn lived down the road from Creeley for a while, in an old adobe that has housed many artists over the years. Bob D'Aessandro, a successful photographer who spent most of his time in New York, used to live out on the San Francisco Ranch road, among an eclectic collection of hand-made houses belonging to many of the long-time Anglo counterculture advocates. Meredith Monk, the dancer and composer, lived there, too. Various other potters, lithographers, musicians, painters, actors, writers, and belly dancers were all a part of the Placitas community, performing at the Thunderbird Bar, building houses, growing food, having kids.

Charlie and Ann Vermont lived down the road from me in the crumbling adobe that would eventually be mine. Charlie and Ann used to run the Boogie Bakery in Albuquerque, where their kids Sam and Sage were born. When their entrepreneurship proved unprofitable and overwhelming, they moved to Placitas for a quieter life. When their quieter life also proved unprofitable Charlie went back to school to become a physician's assistant (he eventually went to medical school and became a doctor) and Ann held down the fort at home. I remember many wonderful evenings at their home, drinking wine, smoking pot, reading poems, arguing politics, meeting all the assortment of people flowing in and out of Placitas. When Ann and the kids left to be with Charlie at his job in Arkansas, my partner Mark and I inherited the house.

I met Mark at the Vermont house. Inspired by Robert Creeley to come out west (Bob was one of Mark's professors at the University of New York at Buffalo) Mark entered graduate school at the University of New Mexico, met Larry Goodell and moved to Placitas. He always said he fell in love with me because I was a fire lookout, but the Vermont's leaving hurried our relationship along as we both wanted their house. Anne Rustebakke was moving back into her Placitas house, which meant evicting me, and Mark lived in John Nordstrom's solar dome, which leaked like a sieve. Not one to choose favorites, Ann Vermont gave it to us both. I guess we have her to thank for our thirty-four year relationship.

Mark and I spent five years in the village working constantly on the leaky roof, clogged pipes, and disintegrating septic line of our hundred-year old house. We also irrigated Anne Rustebakke's orchard, started our own garden, and raised some chickens and goats. The village of Placitas has an extensive *acequia* (irrigation) system fed by one of the spring-fed ponds above the village. Mark and I knew nothing of irrigation—as children we turned on the tap or the garden hose and that's as much as we knew about where water came from—but quickly learned about too much water flooding the lettuce, too little water when your neighbor takes it first out of turn, and gophers, who colonized Anne's orchard. We also found out about how you keep the waters flowing, that annual rite of spring called *la limpia*, ditch day, the cleaning of the *acequias*.

A Saturday in April was the designated day for this Placitas rite. At 7 a.m. the water users congregated at El Oso Reservoir. Most of the grant families were represented, along with the fifteen or twenty Anglo families who own irrigation rights, or *derechos*. The village water system has always been the domain of the Hispano community, although every household with a water right must participate in ditch cleaning or hire someone to participate in its stead. And until the year I dug ditches, the water system had always been a male domain as well. As long as anyone could remember it had been only men who gathered at El Oso to wield their shovels against the weeds and dirt clogging the *acequias*. I certainly never thought of ditch day as a vehicle for making a statement regarding a woman's status in village life. Sometimes necessity dictates the order of things, though, and being an able-bodied person with a *derecho* to maintain (Mark was digging for Anne Rustebakke's two *derechos*), I decided that, barring any furor, I should do the digging for our household.

That year there was a new *mayordomo*, or ditch boss, replacing Jocko, the *mayordomo* of twenty-years duration, who was ill. I decided my first move was to meet this man, test his reaction, and hopefully gain his approval. I found Tony DeLara cleaning the remains of a butchered cow out of the back of his pickup.

"Well, now, so you want to clean the *acequias*. That's pretty hard work, you know. You pretty strong?" I assured him I was, hoping his interpretation of strong was the same as mine. "Well, as long as you're pretty strong I don't see why you shouldn't help us dig."

"I wouldn't want to offend anybody, being a woman and all."

"Yes," he laughed. "I can see you're a woman all right. But don't you worry about anybody else—as long as you do your share, you're welcome."

Well, I did my share and it almost killed me, but then it almost killed some of the men, too. Cleaning ditches is back-breaking work. After an hour and a half at El Oso Reservoir, trying to figure out who was working whose *derecho*, we formed a line at the head of the *acequia,* each *parciante* (water rights owner) assigned an eight-foot length to clean. The object of the endeavor was to clean out all the excess dirt that had caved into the ditch over the winter, pull all the weeds, brambles, grasses, and willows that had taken hold, and generally ensure an even flow of water when the reservoir was opened. As we finished one section, the line would swing in snake-like progression, down the hill to the next section. And so it went, all day long.

Placitas ditch cleaning crew.

I got right into it, anxious to make a good impression, and felt exhausted by the second swing of the snake.

"Hey, man, you're working us like dogs!"

This came from the man working next to me, already done with his section, leaning on his shovel, and was addressed to the *comisionados*, members of the water board, who, along with the *mayordomo*, supervised this activity. My neighbor—dressed in overalls with a red bandana tied around his forehead accentuating the freckles across his nose—was the life of the line. As he moved down the ditch he kept up a steady banter of whining: "Hey, it's too hot to do this, let's go to my house for a drink; hey, man, give this poor lady a break—and me, too, while you're at it." It was his prerogative to complain, I guess, being a member of one of the original families upholding the tradition. I envied his sense of place that allowed him a sense of humor while I maintained an industrious demeanor as I sought to prove myself. And I appreciated that he sometimes dug a few extra feet beyond his section—in my direction.

Around mid-morning, as I followed my part of the line down the hill, everyone started yelling at me that one of the two brothers on the water board wanted me back at the section I had cleaned. Embarrassed, I walked back to where he stood, everyone watching, and was relieved to see him standing over a section that my vociferous neighbor had cleaned.

"This section hasn't been dug deep enough. Water doesn't run uphill," he said to me, very seriously.

"I'm sorry, but that's not the section I cleaned. Mine is the one above that."

Now *he* was embarrassed, and he said gruffly. "Well, who did this section?"

"I don't know—our line got all out of order last time," I said, glad to prove that I could be a good old boy, too, if I wanted.

"Well, Miss, next time don't leave your area until your line has been checked," he said, looking down at his feet.

"No, I won't."

As I walked back to my place, one of the older men winked at me and said, "Did he give you a hard time? Don't pay him no mind. We like having you. He's young—he has to throw his weight around."

The morning dragged on, and my muscles ached with every shovel load, but it was interesting to see the layout of the village as we followed the complex irrigation system down from the reservoir. I saw houses I'd never seen before, and ruins of houses that had probably existed long before the irrigation ditches. Standing above Placitas, looking down, the red-colored mesas of the Rio Grande Valley stretched to the purple outline of the volcanic Jemez Mountains. New Mexico is filled with villages like Placitas whose inhabitants will do the same thing we were doing in a tradition as old and revered. In many ways, Placitas has lost a sense of the past that the other villages still maintain because of their remoteness. Life is different in San Miguel and Truchas, with less opportunity for employment, education, and for government solutions to individual problems. But it is just that isolation that protects a tradition of pride and self-reliance.

At lunchtime, one of the board members discovered that Mark's and my *derecho* was shared by our next-door neighbor, who plowed our back field every year for beans, and that I didn't have to return for the afternoon's labors. Mark gallantly agreed to continue digging Anne's rights (as I hoped he would) despite my gallant offer to help. I went home and didn't move for two hours. When compassion got the better of me I drove down with cold drinks to where the others were working.

Mark

Ditch day ended for everyone late that evening at a party given by the water board members, restoring a tradition that had been unobserved over the last few years. We all drove out to the site of one of the old communes, where a goat was roasting, underground. The men I'd worked with introduced me to their wives, who all congratulated me on my achievement. "I always wanted to help, too," one of them confided to me. "But I was afraid I wasn't strong enough. Maybe I can dig next year." I was relieved they didn't think me an intruder. I dug ditches every spring I lived in the village, and I always felt much the same satisfaction, if not the novelty, of that first spring.

House Building

"To be happy at home is the ultimate aim of all ambition; the end to which every enterprise and labor tends, and of which every desire prompts the prosecution."

—Samuel Johnson

In those halcyon New Mexico days it was possible to make a marginal living with marginal jobs and be quite comfortable. Rent was cheap (we paid $50 a month for our crumbling house in the village), gas was cheap enough for plenty of little trips to Chaco Canyon, Mesa Verde, or the Telluride Jazz Festival, and we took full advantage of the situation by working only enough to pay the bills and have a good time. Mark worked at the Living Batch Bookstore in Albuquerque, while I transferred from the fire lookout on Mount Taylor to the Sandia Ranger District in the Sandia Mountains behind our house. The first year I worked as a hiking patrol up on Sandia Crest, walking the loop between the Crest House and the Sandia Peak Aerial Tramway a couple of times a day, tending to low altitude tourists suddenly overcome by the lack of oxygen at 10,000 feet, La Luz Trail hikers who came up the entire eight miles with no water, and the lost, curious, or friendly wanderers glad to see a woman in Forest Service green. The next year I became the wilderness patrol in the newly designated Sandia Mountain Wilderness and walked the extensive system of trails that cover the mountains, doing light trail maintenance and the public relations work necessary to get the word out that this was now a wilderness area and what exactly that meant. When the fire lookout on Cedro Peak, over in the Manzanita Mountains (still part of Sandia Ranger District), became available I gladly returned to being a lookout, my favorite Forest Service position. It's the only job I know of where one can read *War and Peace*, make a quilt, bake bread, and write stories while collecting an hourly wage.

In 1978 Mark and I began building our own house on land I had acquired soon after I moved to Placitas. Land was still affordable back in those days, too. For a couple of hundred dollars down and a hundred dollars a month, anyone could purchase a home site in one of the little communities surrounding the village. I bought five acres in the Tunnel Springs area, near Cañon Agua Sarca and the North Crest Trail. One of the most beautiful areas in the Sandias, Tunnel Springs was being slowly sold off in five or 10 acre parcels for $1700 an acre. When I purchased my land in 1974, there were two families living in the area, and one in the process of building a dome next to me. The view was expansive, out across the Rio Grande Valley to the mesas of Santa Ana Pueblo and on to the Jemez Mountains. Mount Taylor defined the western horizon; the Sangre de Cristos above Santa Fe loomed to the northeast. Later on, of course, we discovered the drawbacks of living on a mesa top, where the spring winds blow dust in your pores and cobwebs through your mind, and the water is 300-400 feet deep, but, like the later immigrants brought in by the developers, we were overwhelmed by the *ambience*. Mark and I had tried to buy land in the village, because we wanted both the irrigation system that allowed you to grow gardens and trees, and the feeling of community that the village provided on your daily walk to the post office or Lizzie's store. But most of the land was held by the grant and was not for sale.

We continued living in the village for five years before we moved into our house at Tunnel Springs. When you are your own designer, contractor, and laborer—with no building loan—a house does not appear overnight. Just figuring out what we wanted to build took at least a year. We finally decided on the style of northern New Mexico Spanish territorial with solar additions, meaning a two-story adobe with a pitched tin roof and a greenhouse. Each year of building brought certain modifications as we figured out what the hell we were doing. After paying for a well and underground electric lines, it took a year to recoup our losses before we could begin building anything.

Tom Nordstrom, John's son, got us through the first year of the foundation and adobe walls. He, his wife Kimiko, and their two children, Chaim and Naomi, lived in the triple dome up in Dome Valley where they were in the process of adding on adobe rooms in some kind of architectural peace. We couldn't have done it without Tom. Mark met Tom when he was renting the solar dome from John and watched him work on the adobe addition to the domes. When we started building our house, Tom was our part-time laborer, when we could find some money to pay him, and our full-time consultant on how to lay cement block, how to lay adobes, how to build window frames, how to pour bond beams, how to plaster. A small, wiry, intense man, Tom liked to do it all; build adobes; raise goats, free-range chickens, pigs, and turkeys; grow trees, garlic, and vegetables; build ponds and raise fish; sell milk, cheese, eggs, and solar collectors; maintain the life that we came to Placitas to find.

The shell of a house.

We cut the *vigas* over the winter with the help of Jackie, my Forest Service friend and expert tree feller. In return for her aid we helped her tear down a cabin in the mountains and got half the oak flooring. Most of the ceiling went up in the spring, and by summer we were ready to erect our gambrel roof, a big design change from a simple pitched roof to afford us a full second story. Mark refers to it as the fatal decision—he fell off the roof erecting the first gambrel truss and the piece of plywood falling after him split open his head. For a while, they (the doctors in the hospital emergency room) thought he'd broken his neck as well, but it turned out to be a torn vertebral ligament—not as dangerous, but just as painful. He was out of commission for the summer with a very tender head (over a hundred stitches in it) and a very sore neck. I finished off the ceiling and built the rest of the trusses while Mark regained the will to live.

The next year everyone we could corral helped lift the trusses into place. Then it was time for the nailers and tin; Mark refused to climb onto the twenty-foot roof peak to nail the ridge cap (I couldn't blame him), so I did. Tom roofed the dormer windows, hanging out over the windows on a rope.

Other aspects of life were being attended to amidst this preoccupation with building. We both continued working part-time jobs: Mark helped build other people's houses while I worked the fire lookout and for the continuing education program at the University of New Mexico, taking people on hikes in the Sandia Mountains. We planted fruit and shade trees: sweet and sour cherries; pear; plum; apple; peach; willow; cottonwood. After several years of growing a garden at our rented house in the village we decided to work our own ground to establish a good, rich soil. We fought the heat, grasshoppers, no irrigation, and lack of rain to raise vegetables that fed us all summer long. We bought blue grama seed for a natural grass cover around the house, and wildflowers to mix in the meadow. Three dogs, five cats, a goat, and nine chickens rounded out the family.

As I approached thirty another preoccupation rivaled building—babies. No matter how strange it might seem that we would want a complication of that magnitude in the midst of building, we had to insist that there was life beyond building, to maintain the fortitude to *keep* building. Like many of our friends and neighbors we had postponed childrearing until the end of our twenty years' adolescence, and regardless of how complicated or money-less our lives were, it was now or never. I'd always wanted to be a mother, Mark wanted to be a father, our parents wanted to be grandparents, so we did it. Mark began the summer preparing the outside walls of the house for plastering: tacking up styrofoam insulation, covering that with black tar paper, covering that with stucco wire. It was hard work to do alone, and I helped on days off from my Forest Service job. But by July I was rendered useless with nausea. I was pregnant.

When it came time to plaster, in October, Mark was forced to overcome his reluctance to work in high places once we decided it wasn't a great idea for me, five months pregnant, to be on a scaffold. So Mark climbed the three-tier scaffold, I filled buckets with plaster, and he hauled them up. Over the winter Mark worked part-time remodeling a house out on the mesa while I sat home getting fat.

Our son Jakob was born March 5, 1981. Two and a half months later I took him up to the fire lookout with me for my summer job. There we sat, five days a week, in a twelve by twelve-foot tower, looking for fires, changing diapers, walking the twelve by twelve foot floors. Fortunately, Jakob was a calm baby and only occasionally provided background screaming while I was on the two-way radio trying to direct fire crews into the fire. It was a good summer, actually, taking care of Jakob while earning money and assuaging my guilt at not being able to help Mark on the house.

He, meanwhile was in the final stretch (the stretch before moving in, not finishing, as one never finishes a handmade house), helped by our friend Terry, an electrical engineer who preferred building houses. The last year was all interior work: doors, windows, a staircase, plastering, painting, installing plumbing and electricity, laying brick floors, etc. We decided we had to be moved in by fall, or we'd give up and move to New York City and drive cabs. To verify our decision we invited both sides of the family to a reunion-housewarming set for the second week of September. In August, our well went dry.

Placitas is a semiarid environment where farming is viable only with irrigation. The outlaying communities are all served by deep, individual wells, ranging in depth from 200 to 400 feet. Rather than tapping into a water table, these wells are fed by streams of water flowing off the springs and snowmelt of the Sandia Mountains. Our well going dry seemed to confirm our suspicions that Placitas was growing too much and too fast for this water to be replenished, and all of us were going to pay for it. The seventies had brought a new population to Placitas—subdividers. They'd bought up all the available lands surrounding the grant and discovered that they could successfully market the Placitas *ambience* to people who could afford custom adobe homes in "restricted communities," meaning no chickens, junked car-top domes, or dead refrigerators in the backyard. It was inevitable, really; with Albuquerque only twenty-five miles away the area was bound to attract professional people willing to commute to work. Deep, new wells were drilled, roads paved, a real estate company headquartered in the village.

We had to borrow the money from our parents to sink a new well (400 feet deep), but we moved into our house on September 5th, five years after the foundation had been dug. The upstairs was an uninsulated expanse of tin roof, kitchen cabinets were a figment of the imagination, and the wind whistled through unsealed doors and windows. Everyone arrived for the housewarming two weeks later, and we all marveled at how the sun set over Cabezon Peak, far away behind the Rio Grande mesas out our bay window just as the moon rose over the Sandias to the east, illuminating the house against the piñon and juniper trees out our kitchen door. It was ours, for better and for worse, whatever it meant to own something you had created with blood (literally), sweat, and compulsion. By the time we sold our house, in 1992, we had two bathrooms, a greenhouse, a studio, fully mature fruit trees, a garden plot with soil that could rival the Midwest's, a fenced-in backyard with native grass and shade trees, and a newly built breakfast nook plastered in Jemez red mud.

The Sandias

"I do not happen to favor the scarring of a wonderful mountainside just so that we can say we have a skyline drive. It sounds poetical, but it may be an atrocity."

—Harold L. Ickes, former Secretary of the Interior

Placitas lives in the shadow of the Sandia Mountains, the blue jagged backdrop that is the southern boundary of the village. The people of Placitas can walk out their back doors and within minutes be climbing the ridges and arroyos of Cañon Agua Sarca or Cañon del Agua. They have their own routes to the shelved, limestone ridges capping the canyons, the secret caves hidden in the dense thickets of Gambel oak and Apache plume, the ever-flowing Agua Sarca waterfall. And at least on the north end, they're often the only ones there, from the old-time residents to their adolescent grandchildren who think nothing of climbing the ten miles to North Peak and back in a day. They are possessive of the Sandias, these mountains that are so much an integral part of their lives and heritage.

Many people in Albuquerque feel the same way. Minutes away from their homes lie the La Luz, Embudito, and Piedra Lisa trails of the west-side canyons. These people are attached to the mountains, too, but they have to share that attachment with tens of thousands of others. As the population of Albuquerque increases, so do visitors to the Sandias, as more and more people seek recreation in these mountains so accessible to the city.

This accessibility is what makes the Sandias unique. It is good that city people can easily escape the pressures of the civilized world with a visit to the mountains, to experience a forest environment. It balances one's perspective. Yet it is not so good for the mountains' resources and animal populations that must contend with an ever-increasing number of people.

The Sandia Mountains before Albuquerque's northeast heights existed.

In 1978 I was hired to be the patrol for the newly designated 38,000-acre Sandia Mountain Wilderness. I quickly learned where wilderness fell into place in the hierarchy of multiple use. The Sandias are primarily a recreation district, of course, due to their proximity to Albuquerque, but I soon discovered I was the joke of the district. Apparently the Sandias had been designated wilderness over the objection of just about everyone connected with the enforcement of that designation. Some of them were dedicated fire fighters who couldn't imagine walking into a fire with a Pulaski rather than driving there with a chainsaw. Some were dedicated recreationists who couldn't imagine accommodating deer and skunks before dirt bikes and downhill skiing. All were flabbergasted at the idea of a wilderness sitting within ten miles of a large metropolitan area, whose people considered these mountains their own personal playground. I'll never forget the day I was leading a day hike down North Crest Trail with my Continuing Education class and came to the north Agua Sarca overlook to find two four-wheel drive trucks parked in the middle of the trail. They had managed to negotiate the old road cut that leads north from the mountains' paved highway, the ugly remains of a 1960s' ill-fated Forest Service plot to build a scenic loop drive across the north end of the Sandias to Placitas (which fortunately died an ignoble death in the 1970s during the so-called energy crisis). This group of about eight people had also managed to ignore the wilderness signs I had spent a good part of the summer posting at the various boundaries where people were likely to enter the wilderness. I immediately lost my temper, of course, demanded to see a driver's license so I could report this transgression to the Forest Service law enforcement officer, and ended up in a shouting match with a man wielding an ax over a pile of wood (I doubt that I would have been so vociferous if I wasn't accompanied by the fifteen people in my hiking group). I eventually got the driver's license number and reported the trespass to the Forest Service, but a violation was never issued.

In an essay called "The West that was, and the West that can be," Dan Flores, a history professor at the University of Montana at the time, has this to say about the concept of wilderness: "Wilderness is certainly the wrong word for what early America was. It's the wrong word because it's Eurocentric and it obscures more than it reveals. What is obscured is that the garden doesn't have to be free of the human touch to still be a garden."[5] (See also William Cronon's essay "The Trouble with Wilderness; or Getting Back to the Wrong Nature" in *Uncommon Ground* for a definitive exploration of the concept of wilderness.) The concept Flores refers to has come to be called "inhabited wilderness" and will be more fully explored in later chapters that deal with the protection of northern New Mexico communities. In a designated wilderness area such as the Sandias—a mountain range completely surrounded by an urban environment—perhaps the discussion should revolve around "what kind" and "how much" of a human touch is appropriate. A lot of rhetoric is bandied about over the issue of wilderness versus developed recreation, including the "elitist" label slapped on wilderness proponents by those who prefer unrestricted recreation in all our national forests. When one considers the level of impact and economics of hiking, camping, and cross-country skiing versus snowmobiling, driving, and alpine skiing that label takes on some irony. The Sandias have always provided enough developed recreation: picnic sites; a scenic drive to a gift shop and magnificent view; an aerial tramway to a restaurant and downhill ski area. Because of their proximity to a city of almost 500,000 people, they can also provide a preserve for plant and animal life, and for the Eurocentric concept of humans as visitors only, experiencing "primitive recreation and solitude."

In Flores article he acknowledges that the Native Americans on this continent "managed" their landscape by clearing forests, draining swamps, and diverting water. They may also have played a role in the extinctions of species. But they numbered only 10 million people at the time of European contact and their land management enabled their sustainability. The "managed" landscape that resulted over the next 250 years is why an area like the Sandias is such an important preserve: it is as close as we can come to "untrammeled." Unfortunately, because the management of these landscapes has included the suppression of fire, one of the ways the Indians "cleared" the forest, the Sandias are also a tinderbox of overgrown mixed conifer and underbrush that could carry a conflagration through every other manmade development in a matter of hours.

By building blacktop highways, alpine ski areas, restaurants, and tramways the Forest Service does little to sensitize people to issues concerning wilderness and conservation. Because so many of us now inhabit cities rather than forests, and our only contact with nature is as a visitor, there is not much understanding of the concept of inhabited wilderness, where one is connected to the land through place and work. But when you're out there every day, walking the trails, climbing the granite boulders, listening to the wind and rain roll rocks down the face of the mountain, as people have for millions of years, perceptions change. When you come to know that the delicate purple and yellow flower you see growing along North Crest Trail in early summer is a lady slipper orchid, and the plaintive call coming from the thick spruce-fir forest is a Townsend's solitaire, you gain a connection to these mountains which expands your world beyond your own often nebulous and fragmented concerns.

Marketing a Way of Life

"We call ourselves hippies with beepers."

—Tom Ashe, Placitas developer

While John Nordstrom and Ralph Roller could be considered the first of the Placitas subdividers, it wasn't until the late 1960s that the real land grab began. Several Anglo families bought chunks of land around the village for a small percentage of their eventual worth and began subdividing the property. The McKinnon family (the same family who ran the restaurant where Lizzie Archibecque worked) developed what was to become one of the largest subdivisions in the area, Ranchos de Placitas. The McCallister family bought up land in the Las Huertas Valley between the national forest and the village and built suburban houses of ranch-style frame and brick, quickly dubbed Gringo Gulch. Rumblings of discontent erupted into direct action when the sales-office trailer in Ranchos de Placitas was bashed and real estate signs were burned and defaced. These were spontaneous acts that reflected local resentment toward the more affluent people moving into these developments.

That was only the beginning. In 1981 two Placitas couples, Steve and Wendy Gudelj and Rick and Pepi Levin, formed the Placitas Land Co., a real estate firm. About the same time, the McCallister family subdivided a piece of land called Placitas Homesteads about three miles west of the village. A year or two later, Rick Levin, Steve Gudelj, and a third Placitas resident, Tom Ashe (married to Pepi Levin's sister Joanne), started Placitas Trails, just west of Homesteads. Norman Lazar developed an area called Terra Madre, with Levin and Gudelj as partners.

The developers restricted the three subdivisions to ensure a suburban nature: covenants excluded farm animals, mobile homes, and multifamily dwellings (although townhouses were built on part of one of the subdivisions). Most of the lots were between 1.7 and 2.2 acres, at a cost of $17,000 to

$25,000 an acre (they would later go for $60,000 or more an acre). Two of the developments, Placitas Trails and Tierra Madre, allowed houses no smaller than 1,700 square feet. Various contractors built houses in the developments (Tom Ashe was one of the busiest), but the architectural styles remained similar: sprawling frame-stucco with passive solar features.

The Placitas Land Company marketed Placitas as the ideal community for the upwardly mobile professional seeking an alternative to the urban environment of Albuquerque. One of its radio spots described Placitas as "what Santa Fe used to be and Albuquerque never was." Other ads presented Placitas as a unique community of artists, farmers, and professionals with expansive vistas and access to the national forest—the New Mexico adobe tradition in laid-back rural style.

The developers marketed *themselves* as Placitas saviors. In an article I wrote for the *Albuquerque Journal* in 1987 (I began to write freelance articles for various publications in the early 80s), I interviewed Rick Levin and Steve Gudelj. Referring to their Placitas Trails subdivision, Levin insisted that he and Gudelj were saving us: when they heard that a developer was planning to turn the area into a mobile home park "we decided to see if we could purchase that property and turn it into a quality residential subdivision. It was a way to do something about the inevitable development of the area. The mobile home park would have been inharmonious to what we felt was going on in the area."[6] It's a tactic often used in defense of an unpopular act—if we don't do it, someone else will. The Placitas Land Co. used mobile homes as their scapegoat. Everyone loves to hate trailers, despite the fact that they are some of the only affordable housing available to people. But developers don't make money off affordable housing. They make money off the $500,000 homes that are built in their subdivisions.

Witnessing the success of Placitas Land Co., established developers and contractors from Albuquerque moved their operations to Placitas and gave us Vista de Montaña, La Mesa, Juniper Hills, etc. But the unkindest acts of all, in the minds of many, were when some of the local folks decided to cash in. One of these people, Bob Poling, had originally come to Placitas for the same reasons I did, or so I thought. At least he partied with the same people I did, built his own house, started up a small cottage industry to make a living. He eventually became a partner with Ashe and Gudelj, then a developer on his own in Cedar Creek, northwest of the village on lower Las Huertas Creek, and the Overlook, on an escarpment above the village. Others became successful contractors, building luxury homes for Poling, Gudelj, and Levin, while still others made a good living building their cabinets, contracting their plumbing and electrical work, and plastering their fake adobe walls. While we tried to make the distinction between the "workers" (very few of them were from the Hispano community) and the "developers," it was hard to not be judgmental.

Signs of discontent by those unwilling to cash in again appeared in the eighties. One morning we woke up to find that the Placitas Land Co. signs, which featured a black buffalo, had been altered to depict a white buffalo mounting the original one. The company offered a reward for information about who had defaced the signs, but no culprit was publicly identified. In my *Journal* article Rick Levin said, "We know who did it. It was a statement of unhappiness by a few people who thought it would be a great joke. And we appreciated it in terms of a joke. It also expressed the sentiments of a very, very small number of people out of the total population that lives here."[7]

Another creative act soon added to the merriment. Newsletters entitled "Placitas Unreal Estate News" were mailed to various Placitas folks, parodying the Placitas Land Co. newsletter, which was periodically mailed to the entire community. Following are excerpts from the *Unreal Estate News*:[8]

Any Hill Left Unscarred? . . . We'll Adorn It With $200,000 Boxes
Any Jackrabbit Left Roaming? . . . We'll Take Away It's Habitat and
    Give It to the Dogs
Any Rural Diversity Left? . . . We'll Suburbanize It With Sameness

Businesses expected to be seen in the strip shopping malls along SH
165:

   *Holistic Public Accountant
   *No Appointment Liposuction
   *Laughing Boor Art Gallery and Pine Tar Sachet Shop
   *Homeopathic BMW and Mercedes Benz Car Maintenance
   *Community Outreach Gun and Burglar Alarm Boutique
   *Love Thy Neighbor Motel

*Two Minute Organic Food and Nonalcoholic Beverage Services
For Those Too Busy To Do Anything Other Than Make Money

Renaming the Subdivisions:

*Cedar Crook
*Overkill
*Ashehole Flats
*Bumsteads
*Tierra Chingada
*Ranchos de Placebo
*Placitas Depths
*Vista de Nouveau Riche

The land company's response to these newsletters was the same—moral outrage and indignation that a "small segment" of the community was making trouble for these pillars of the community. Word got around they were threatening to sue for libel if they could ever find anybody to sue. Lynn Montgomery got hold of the "Unreal Estate Newsletter" and enjoyed it so much he had the "glorious rag"[9] reprinted and mailed to every box holder in Placitas.

In July of 1988 *Albuquerque Living* magazine ran a very slick article called "The Placitas Good Life"—which showed not the "collection of school buses, teepees or underground homes, the Placitas trademarks of the '60 and '70s"—but extolled the "new" life full of beautiful homes, swimming pools, and colorful gardens. Sidebars featured a multitude of things to do in Placitas: arts and crafts fairs, fiestas, parades, etc.[10] Some folks* decided to let everyone know that the "old" Placitas did indeed exist, much to the chagrin of the "new" Placitas, and produced a brochure entitled "Placitas: The High Life—There's less to this town than meets the eye." Parodying the format of the *Albuquerque Living* article, the brochure described the population as a "number of relatively normal folks...along with alternative agriculture advocates, burned-out heishi [shells used to make necklaces] makers, free-lance auto mechanics, massage therapists, starry-eyed aura balancers, and a variety of kinky metaphysicians, cranks, and other assorted weirdos."

Housing was described as an "eclectic assortment of underground houses, domes constructed of brightly colored junked car tops, school buses, and railroad cars," surrounded by "duck ponds choked with algae, lush lawns of ragweed and goatheads."

Placitas events and activities were listed in side bars: "The Placitas Fire Brigade responds to the torching of the Thunderbird Bar and the demise of the freeform heishi factory, both historic landmarks of Placitas hippie culture;" "The community gets together once a year to boo and hiss the politicos and developers who want to turn it into a rich suburb of Albuquerque and make lots of money."

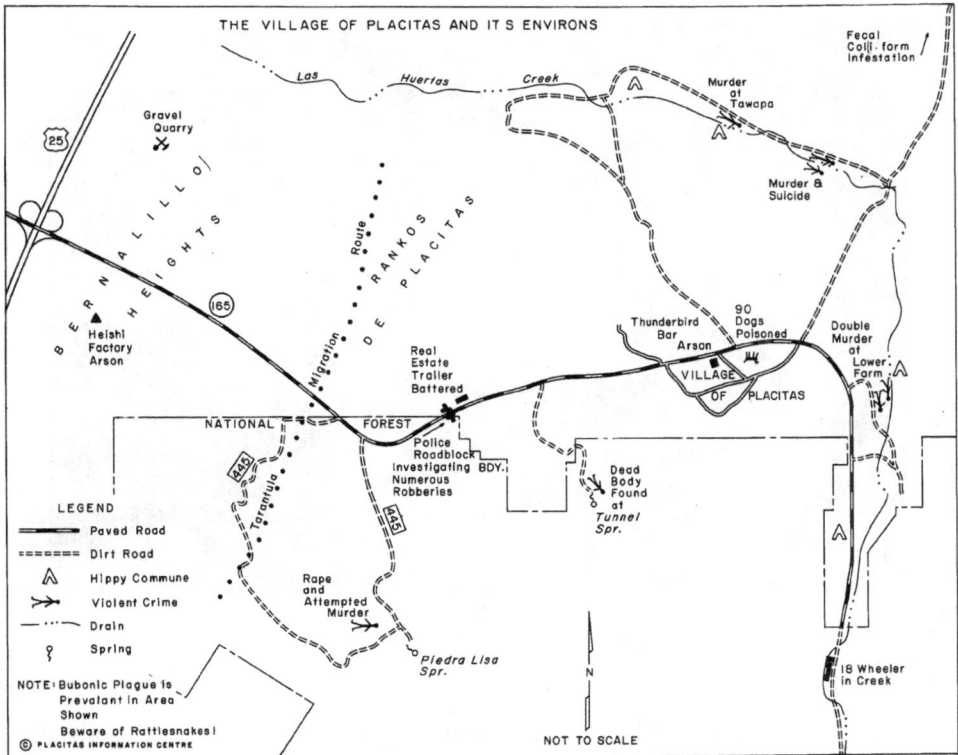

Placitas map by Lenore Goodell.

A map of the area was included in the brochure identifying many famous landmarks: the fecal coliform infestation out on the north mesa; the tarantula migration route; the site of the infamous dog poisoning that killed 90 dogs; the Towapa murders. Black and white photos also illustrated the article. A picture of a trash heap in an arroyo was captioned: "Native landscaping can also be used imaginatively in arroyos, making trips to the dump unnecessary. Here, an old freezer, some Tonka trucks, and pieces of chicken wire blend gracefully with the native tumble weed and juniper."[11]

Other acts were more blatant. The trailer that advertised in huge letters on its side for the Placitas Acres subdivision was rewritten to read, "Placitas Aches." The entrance sign to Placitas Homesteads was amended to read "Placitas Homesteads sucks and so does the PLC [Placitas Land Co.]," and a street sign in Tierra Madre became "Tierra Raper."

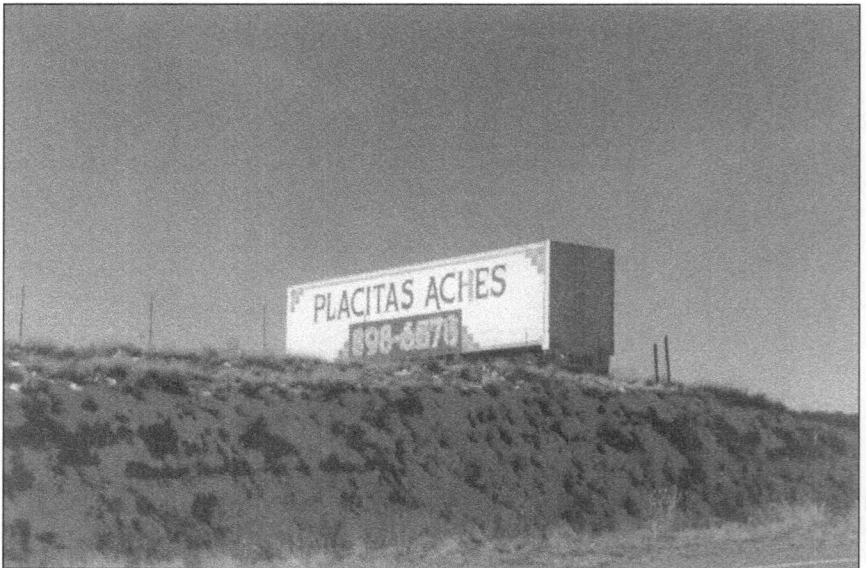

Placitas aches.

The struggle over land use in Placitas is a microcosm of land struggles all over New Mexico between indigenous people and those who wish to profit from control of their land. As rural life becomes more difficult to pursue, as the economy becomes more specialized and less diversified, agricultural land falls prey to development, both industrial and residential. Here in New Mexico, as our economy becomes more and more dependent upon tourism, alfalfa fields eventually give way to homes as subdividers buy up all available land to convert to expensive residential areas, catering to the wealthy and elite. In collusion with the developers are government land planners, zoning boards, and town councils, anxious to raise the tax base to fund their services.

The struggle to control Placitas, where land prices were obviously on the rise, was played out in all the various boardrooms of Sandoval County during the '70s and '80s. Zoning, one of the biggest battles, clearly revealed the inadequacies of governmental attempts to control land use.

* The same folks who published "Placitas: The High Life" published the "Unreal Estate Newsletter." Their names: John Kennedy, Mark Schiller, and Kay Matthews.

To Zone or Not to Zone

"The bulk of rural land use policies...confine growth to paths that do not help the poor much. They protect the environment for people securely established in it and stimulate economic development for people who already profit from it."
　　　　　　　　　　　　　—Frank Popper, Rutgers University

In the early 1970s, several longtime Anglo residents started a petition to declare Placitas a "special zoning district" and incorporate the village. These people were primarily motivated by the desire to gain control of the water board—which oversees the domestic and irrigation water systems— the only organized governing body in the village. There had already been battles over control of the water, breaking down over ethnic lines, and the Hispano population was immediately wary of the move to incorporate. Such a move would also endanger the grant: an ordinance setting a minimum land size, which is almost always included in zoning regulations, would make it difficult for heirs to inherit small parcels of land.

Other residents, including many of the younger immigrants, thought zoning was a threat to their way of life, the do-your-own-thing nature of the community. John Nordstrom told me, "I knew it was an attempt by some of the Anglos looking to make money by getting control of the village and increasing land values."[12] Villagers challenged the zoning request, and the petition was withdrawn.

Zoning again became an issue in 1985 when a family living in an older subdivision, the San Francisco development, appeared before the Sandoval County Commission to complain about a wrecking yard that had moved next door. The family asked the commission to consider a zoning ordinance to protect their residential community from what it considered additional eyesores.

A few months later Steve Gudelj, Rick Levin, and Tom Ashe, the developers of Placitas Trails, attempted to sell nearly five acres of land in the subdivision to a horse stable. The property, once a landfill, could not be used for houses. In my 1987 *Albuquerque Journal* article, then-Sandoval County Commission Chairman Pete Salazar, a resident of nearby Placitas Homesteads, had this to say about the developers' plan: "When I first heard about it I thought it would be a stable for the residents of Trails. Then we found out that the stable could house up to 150 horses, have a training arena and managers' living quarters, and some of the neighbors began to get pretty concerned about the traffic problems, health problems, that kind of thing."[13]

Although the stable developer withdrew his offer, the County Commission passed a resolution asking the Bernalillo Town Council (Bernalillo, situated on the Rio Grande, is the Sandoval County seat and closest incorporated town to Placitas) to extend the town's extraterritorial planning and zoning jurisdiction 2.5 miles beyond its municipal limits. The city and county eventually entered into a joint powers agreement and established a Greater Bernalillo Extraterritorial Zoning Authority and a seventeen-member commission to develop a county plan as well as a zoning ordinance for the area. The authority would eventually be only one of several simultaneous—and sometimes confusing—efforts to bring zoning to Sandoval County.

But it soon became apparent that, once again, not everyone was in favor of zoning. Residents of Placitas West, one of the older, unrestricted communities within the authority's jurisdiction, protested the move at a hearing in Bernalillo. They weren't sure how zoning would affect their community, with its chickens, goats, and trailers. In addition, several people from Placitas wondered how the issue had ended up with the town of Bernalillo instead of the county. And the subdividers didn't want Bernalillo telling them how to run their developments. According to Pete Salazar, "I ended up taking a lot of heat from the realtors when we formed the joint powers agreement to set up the Greater Bernalillo Extraterritorial Zoning Authority. But that was the best thing the county could do as an interim measure to get some kind of control until we could get moving on our own zoning authority."[14]

The county began advertising for a consultant to help develop a county plan and zoning ordinance. But finding one took more than a year. In the meantime, several other land-use controversies arose that intensified the zoning issue.

In the spring of 1985, the city of Albuquerque sought bids on a 980-acre tract bordering Ranchos de Placitas. Albuquerque had acquired the land in a land exchange with the Forest Service. Several people—including an aerospace manufacturer who was contemplating building a plant in the area, and a partnership consisting of Placitas lawyer John Kelly and developer Steve Gudelj—bid on the property.

Although all bidders eventually indicated they planned to build houses on the land, Salazar said he received numerous phone calls from local residents worried about the sale of the isolated tract since Sandoval County had no zoning ordinances to control development. Many of the people nearby in Tierra Madre and Placitas Trails were worried that a huge new subdivision would impact *their* subdivision by creating more congestion and lowering their land values.

Another controversy arose when then-New Mexico Attorney General Paul Bardacke said John Kelly's role as Sandoval County attorney as well as bidder on the Placitas property had the "appearance of impropriety" since Kelly would eventually advise the County Commission about zoning issues in the area. In defense of his position, Kelly said, "As long ago as 1982, when I sat on the County Commission, I encouraged the adoption of a zoning ordinance, the only effective tool to give public and county officials a say in land-use planning."[15] Privately, when I interviewed him for the *Journal* article and asked if he felt conflicted, as a longtime resident of the Placitas village, about joining forces with the developers, Kelly said to me, "This is strictly a business decision."[16] Kelly later resigned his county post after Salazar criticized him for failing to inform county officials that the parcel was for sale. After Albuquerque accepted his and Gudelj's bid of $2.3 million for the property, Kelly and his family moved to Albuquerque. Today his subdivision, La Mesa, extends for miles through the canyons and over the mesas of the Placitas foothills, full of expensive homes and commuter residents. Kelly himself, an old school friend of Bill Clinton, subsequently became the U. S. Attorney for New Mexico.

Controversy continued. In the summer of 1987 Centex announced plans to build a $20 million wallboard plant in Algodones, a community north of Bernalillo along the Rio Grande; and the people living in Tierra Madre subdivision learned that the Springer Building Materials Corp. had leased 1,000 acres of land adjacent to their subdivision for a gravel pit.

In response the Sandoval County Commission established a 15-member County Planning and Zoning Task Force to get the gears in motion for eventual countywide zoning.

"We're mostly concerned with the village, but we own lands scattered all around that would be affected by zoning," Lizzie Archibecque later said at one of the many meetings regarding the zoning issue. "We want our families to be able to do anything they want with their land. I get along fine with everyone, including people from Ranchos and Homesteads, and I don't want to tell them what to do. But I think we in the village—even from Tunnel Springs east up into the canyon—want to be left alone. When you've been here as long as we have, those people who only stay three or four years, or even 10, don't compare."[17]

In preparing my zoning article for the *Journal* I interviewed a university professor whose work in land planning issues I found to be progressive and pertinent to the issues facing Placitas. Frank Popper, today famous for his plan to revert farmland in the Midwest to its original native grasses, was a professor of urban studies at Rutgers University. According to Popper, rural zoning is often the product of urban elites who have moved to the countryside and want to protect themselves and their surroundings. It tends to stimulate economic development for people who already profit from it.

"Urban-style state and local programs to limit the density of rural land use frequently favor large developers, make land markets rigid, increase costs of builders and buyers, support or improve property values, and even promote economic development," Popper has written. That's why, he said, you so often see developers sitting on zoning authorities: "If zoning is supposed to control development, it's developers who eventually control zoning."

Zoning is a uniquely American concept. It developed in the Northeast in the late 19th and early 20th centuries in response to widespread concern over the adverse consequences of rapid growth in metropolitan areas. However, Popper says, early zoning regulations were also aimed at segregating poor people. For example, New York City's landmark zoning ordinance, passed in 1916, fought both unpopular land use and unpopular poor people. "Wealthy Protestants and German Jews on Fifth Avenue instigated it to protect their affluent residential and shopping district from the spread of Seventh Avenue garment factories owned by poorer Polish and Russian Jews and operated by far poorer immigrants from a variety of ethnic backgrounds," Popper writes.

He contends that zoning is a uniquely urban concept that does not travel well to rural areas.[18]

The developers all responded to Popper's contentions with predictable platitudes. John Kelly insisted that development was inevitable and that zoning was the only way to control it, that "most people would much rather see a quality residential development like we propose than the possibility of a commercial development...people's objections to houses being put on our land come from purely self-interest. No one has the God-given right to keep the land the way it is."[19] Steve Gudelj, while admitting that his developments had accelerated growth in the area, stated that most of the people in the community think his subdivisions are nice: "I feel that as a developer I have provided a service to the community with quality housing. Placitas Land Co. works with people from the community to help them find what they want and protect their interests as homeowners."[20]

The Placitas Land Co. always tried to portray itself as a community oriented business and its realtors and builders as solid citizens of the community. Again and again they stated that in the Placitas area development "was inevitable," and they—as residents of the community, whose kids attended the Placitas school, who "can't even afford to live in my own developments" (Steve Gudelj)[21]—were saving us from trailer parks and unwanted commercial development with $300,000 "quality housing." As Tom Ashe put it so succinctly in an interview in the *Albuquerque Journal*, they were just "hippies with beepers."[22] By the early 90s, however, almost all of them would have made their money and be gone to greener pastures, leaving their legacy for the rest of us to live with.

At War

"Science in its proper place, as a servant only, ought to have placed material in ever more copious and diverse supply at the disposal of the poet, who would have made use of it in such a manner that man would have grown up without turning into a monster; but science, as a master, has turned man into a monster, upset the balance of the world, and enabled us to take our commodity from the planet, but to cease giving to it."

—Edward Hyams, *Soil and Civilization*

By 1985 Placitas was completely under siege. Placitas Land Co. was firmly entrenched, making the kinds of profits that Fortune Magazine gives awards for, and Sandoval County government well along the path towards eventual countywide zoning, basically in collusion with the developers. Nineteen eighty-five was also the year the Cibola National Forest Environmental Impact Statement was released, opening the door to further development of the Sandia Mountains and Placitas environment.

The National Forest Management Act of 1976 mandated that all national forests release 10-year management plans (with 50 year guidelines) to govern all facets of its multiple use policies. The Cibola National Forest Environmental Impact Statement (EIS) was one of the first plans released, and saw little of the ground work that became *de rigueur* after so many plans were appealed in ensuing years—consultation with community and environmental groups and pre-draft negotiation.

The Cibola Plan was a compendium of an environmentalist's worst nightmare, including a 300 percent increase in timber harvest levels by the 1990s; consideration of an additional alpine ski area in the Sandia Mountains and another under consideration on Mount Taylor; inadequate protection of water resources; inadequate rangeland protection and permit system; failure to protect the forest's cultural resources of prehistoric and Indian religious sites; and the proposal of a paved road through Las Huertas

Canyon above Placitas. Marion Davidson, as president of the Sandoval County Environmental Action Community, and longtime community activist, might have despaired over such a plan, but managed to channel her energy to anger and activity.

Marion Davidson

Marion lived in what is called Rainbow Valley, in a rambling house up against Forest Service land. She shared her life with numerous dogs, cats, chickens, and geese that I had to fend off with a stick whenever I took care of her animals so she could go off on various treks in the Sierra Nevada, where she grew up. Marion learned her skills on the battle lines of the Student Nonviolent Coordinating Committee back in the early '60s, and as a lawyer out on the Navajo Reservation in Arizona, where she fought the poverty and paternalism perpetrated against our nation's largest group of Native Americans. Many of her Diné friends still came by to visit her in Placitas, to take a sweat bath and share a feast, maybe cross-country ski a few miles into Las Huertas Canyon, while Marion plotted the next step in the long fight over the canyon. Marion was a strategist, but she was also in it for the long run and stayed in Placitas until the bitter end, hanging onto her land and lifestyle, directing the action. Finally, in 1998, she gave up, too, and moved to Oregon to live near her family.

Cibola National Forest is composed of four far-flung ranger districts bordering both urban and rural communities in central New Mexico. The smallest district, Sandia, governs the Sandia and Manzanita mountains, Albuquerque's backyard. Mount Taylor District, where I worked the fire lookout, ranges from the low-country Zuni Mountains, checker boarded with private ranch lands, to the alpine terrain of Taylor and Mosca peaks, center of the universe. Mountainair and Magdalena districts are remote, lesser-used forests inhabited by Hispano and Anglo ranchers and farmers who have lived there for centuries. A broad coalition of groups from these communities—an association of land grants, the Sierra Club, a Native American environmental group, the New Mexico Wildlife Federation, an irrigation ditch association, and a Placitas community action group (of which I was a member)—joined to file an appeal of the Cibola Forest Plan, based on two major issues: 1) the plan's failure to recognize the needs and interests of the forest's Native

American and Hispano communities; and 2) its emphasis upon resource consumption and commodity production at the expense of natural resources.

The coalition filed an administrative appeal of the Cibola Forest Plan in 1986. The development of national forest management plans every 10 years that "provide for public participation in the development, review, and revision" of these plans was stipulated by the National Forest Management Act (NFMA) of 1976, and administrative appeals of those plans were built into the National Environmental Policy Act (NEPA). Both are supposed vehicles for the public to have more input in management decisions. Each Forest Service decision made under the auspices of an Environmental Analysis or Environmental Impact Statement can be administratively appealed by citizens, within a framework of rules and regulations stipulated by NEPA. Seeking to avoid redrafting the plan or a battle in court, the Forest Service offered to negotiate an amendment to the existing Cibola plan. While there was a lot of initial disagreement among the appellants over whether we should negotiate with the Forest Service, we finally agreed to the amendment with the understanding that we could always resume the appeals process if we didn't gain enough concessions in the amended plan.

For six months representatives of the appellant groups and Cibola National Forest sat around a table at the Natural History Museum in Albuquerque arguing over timber dollars versus trail building, cows versus watershed plans, cross-country ski trails opposed to alpine ski slopes, twice a week, every week from May to November of 1986. The appellants saw our job as twofold—to get the Forest Service to favor "dispersed" or low-impact recreation in the Sandias, besieged by the half million Albuquerque population at their base, and to reduce commodity management in other areas of the Cibola to protect the needs of land-based communities and the natural resources of the forest. We were asking that the Forest Service leave the back rooms of the timber industry and implement a *real* multiple use policy. Recreation issues centered on the possibility of additional alpine ski area development in both the Sandias and on the slopes of Mount Taylor, an utterly ridiculous idea. None of us wanted a ski area on this sacred mountain.

The best we could do for the Sandias was to get the Forest Service to agree that there would be no consideration of a new alpine ski area until the second decade's planning process (which began in 1995). This issue was a real stumbling block for me. I couldn't comprehend what rationale the Forest Service possibly had for a development of that magnitude. The Sandias were already home to the Sandia Peak Ski Basin, which was a fairly marginal area, due to its poor placement on the mountain and a succession of bad snow years. On weekends and holidays traffic congestion from the ski basin choked the mountain highway leading to the ski slopes, while cars, after filling up the parking lot, snaked along the road shoulder for miles up the mountain. Where they ended was the beginning of the haphazardly parked cars of the cross-country skiers, who found every available parking spot from the ski basin to Sandia Crest. And on top of that were the "tubers," who not only filled Capulin Snow Play Area to capacity but illegally rode their tubes down every available hill *below* the ski basin. It was a logistical nightmare for the Forest Service, yet here they were proposing the possibility of an additional ski area in upper Las Huertas Canyon, accessed off the same paved road. As I sat through the negotiations, it finally became evident to me that developed recreation, as opposed to dispersed, or low impact recreation was the Forest Service choice because the more money the district spent, the more money it received from the budget pie. But that was not the only factor involved in the ski area proposal. To understand it completely one has to understand Bob Cooper, the man behind the scene, but I never could. Bob was an enigma to me.

Bob Cooper

Bob Cooper owned the old Ellis Homestead in beautiful upper Las Huertas Canyon, where a series of seven springs have created a riparian oasis of trout ponds and travertine formations. Bob and his family lived in the original Ellis log cabin during the summer months, and lovingly kept the cabin in its original state, sans electricity and the rest of our 20th century amenities. Bob gave me a tour one day, while I was on patrol in the canyon. A fellow Forest Service employee, I had known Bob casually for years, mostly as a crusty voice coming over the airwaves on his own patrol route on the other side of the mountain.

Bob showed me the cabin first, with its iron bedsteads, wood stoves, and chinked log walls. Then he took me down to the springs that feed Las Huertas Creek, where we walked along the series of pools he had created and stocked with trout to create a riparian area of unparalleled beauty. We finished the tour at old man Ellis's grave, lovingly attended to by Bob, who had lived with the Ellis family as a boy and had inherited this pristine piece of property in the middle of the national forest, the jewel of the Sandias.

Bob wanted to turn his Ellis homestead into the base support for a ski area in the upper canyon. He'd been promoting it as such since before Sandia Peak Ski Basin existed, and been shouting "I told you so" about the area's bad conditions ever since the ski basin owners chose the wrong site instead of *Bob's* site. After years of working on the Sandia District, he went to work for the Santa Fe Ski Basin, to exercise this ski bug that apparently has hold of his psyche. I've never been able to figure out how a man who so carefully preserved one of the most beautiful places in the Sandias could continue to lobby for a downhill ski area in his backyard. Bob was also the main advocate of paving the Las Huertas Canyon Road, to enable those potential skiers to reach his development (and the various other cabins on the Ellis Homestead, developed by Paloma Park, one of Bob's business partners). The next chapter discusses that battle, which polarized the Placitas community and left Bob and me adversaries instead of friends. Sadly enough, the Ellis cabin was burned down by arsonists in 1991. Bob's first statement to the press was that if the road had been paved fire fighters could have reached the site more quickly and perhaps saved the cabin.

Many issues in the Cibola Forest amendment remained unresolved. Most of them revolved around timber—below cost timber sales, timbering in semi-primitive non-motorized lands, old growth protection—grazing, and American Indian religious rights. The appellants decided to sign off on the areas of agreement and continue with the appeal of these unresolved issues. But the Forest Service refused to sign the agreement unless we dropped the appeal. We argued long and hard among ourselves, feeling this preliminary sign of bad faith didn't bode well. In the end, however, we agreed to their conditions, knowing that we could always appeal individual Forest Service decisions at a later date.

Unfortunately, that's what happened. Since the release of the Cibola amended plan in 1986, there has been an appeal of a timber sale on Mount Taylor (which did not resolve any of the timbering issues raised in the negotiation of the plan), an appeal and lawsuit of the forest's decision to develop Las Huertas Canyon, numerous incidents of noncompliance with the agreements made in the plan, and the implementation of projects that were never even *included* in the plan. While the agency was careful to notify citizens of planning sessions and included them on field trips to issue sites, the working sessions I attended broke down in discord. We were faced with a never-ending string of appeals as each management decision was implemented as the Forest Service saw fit rather than what was called for in the amendment. At the time, I was distressed that we weren't tougher in the Cibola, that while we negotiated in good faith, perhaps we let down our guard and they cut the coyote tree—Forest Service lingo for the large, seed-bearing tree they leave standing—right in front of our faces. The following chapter demonstrates, in excruciating detail, just how meaningless NFMA and NEPA have been in guaranteeing any real citizen input in the way our forests have been managed.

The Fight for a Canyon

"Recreational development is not a job of building roads into lovely country but of building receptivity into the still unlovely mind."
—Aldo Leopold

Las Huertas (the gardens) Canyon lies on the east side of the Sandia Mountains. Through this canyon a narrow dirt road winds eight miles down from the alpine forests near the crest of the mountains to the piñon/juniper valley of Placitas. Alongside four miles of this dirt road run the cold waters of Las Huertas Creek, one of only two perennial streams in the Sandias.

For almost 20 years, local debate raged over the fate of Las Huertas Canyon. In its 1975 Sandia Mountains Land Management Plan, Cibola National Forest delineated the problems that plagued the canyon: sedimentation of the creek; traffic congestion and hazards to public safety; noise and dust pollution; littering and vandalism; destruction of the wildlife habitat and stream-side vegetation. Added to that, the Sandia Mountain Wilderness, designated in 1978, borders the road to within 100 feet, and folks driving into the wilderness was common.

Placitas is at the north end of the canyon and many of its residents have long used and enjoyed Las Huertas. Land grant heirs consider the canyon theirs; this was their commons land, where they hunted, grazed their sheep, and cut their firewood. Because it is literally in the village's backyard, the people of Placitas have a special interest in protecting the natural resources there. However, many Albuquerque drivers, enjoying all the Sandias have to offer—the aerial tramway, picnicking, restaurants, skiing, and hiking—used Las Huertas Canyon to complete a loop drive around the mountains. On weekends and holidays, increased traffic in the canyon created bottlenecks, caused clouds of dust, and made it unsafe for people to hike or fish alongside the road. The residents of Placitas had to contend with speeding traffic, and the volunteer fire and rescue brigade had to respond to a high incidence of accidents.

Since the release of the 1975 management plan, the Forest Service had at various times proposed three basic management alternatives to alleviate the problems in the canyon: 1) paving and reconstruction of the road to a 35-mph, two-lane thoroughfare; 2) rerouting the road out of the canyon (it currently has a state highway designation, but the Department of Transportation has periodically considered dropping the road from the system); and 3) closing all or portions of the road to vehicular traffic.

The 1975 management plan was scrapped with the inception of the National Forest Management Act (NFMA) in 1976, which, as I described in the previous chapter, required more comprehensive, 10-year plans for all the national forests. The Cibola Management Plan, released in 1985, again proposed the three alternatives of the 1975 plan, and the 1986 negotiated amendment to the plan stipulated that a citizen work group representing affected interests help write an Environmental Impact Statement (EIS) for Las Huertas Canyon.

I served on the committee as a representative of the Placitas community, along with representatives of the San Antonio de Las Huertas Land Grant, The Pueblo of Sandia, Albuquerque, the State Highway Department (now the Department of Transportation), landowners in the Ellis Homestead, Las Huertas-La Jara Ditch Association, and several environmental groups. We listed eight possible management alternatives, which included no action, several kinds of road improvement, and several kinds of road closures. The group majority then made a formal recommendation to the Forest Service that the lower, northern portion of the road, above the village of Placitas and next to the creek be closed, with the remaining southern portion open from the Sandia Crest Highway as access to the private inholding (the Ellis Homestead) and a Forest Service picnic ground.

The group was essentially an exercise in futility. After the work group had spent months formulating the management alternatives, John Daley, of Paloma Park Development Company, land owner along with Bob Cooper of the Ellis Homestead and designated representative to the work group, delivered a letter to the Forest Service stating that landowners regarded *any* partial road closure as a denial of their access rights and they would sue to protect those rights (Daley had not participated in drawing up the management alternatives). Threatened with a lawsuit, the Forest Service announced it would not consider any road closure alternatives.

Typical day in Las Huertas Canyon.

When the Draft Environmental Impact Statement (DEIS) was released to the public in July of 1988, the Forest Service chose to ignore our recommendation (which we knew it would) and proposed paving the entire length of the road. It also rewrote much of the group's writing of the eight alternatives, substantially changing the spirit and intent of the proposals. As a rationale for choosing road development, the Forest Service stated in the DEIS that paving provides for the "highest level of additional motorized opportunities" and the "highest level of additional recreation capacity." The agency contended that though paving might exacerbate some of the existing problems in the canyon, it was an environmentally sound proposal—despite the fact that in the DEIS the Forest Service chose one of the road-closure alternatives as its "environmentally preferred" alternative.

True or not, the specter of a back room deal where the developer and the Forest Service were the only participants loomed ominously over the proceedings. Bob Cooper was already on record as advocating paving of the road to facilitate development of a ski area in Las Huertas Canyon. Paloma Park Development Company was already operating a trout fishing operation and Christmas tree farm on their private land, and John Daley also ran an industrial and commercial real estate office in Albuquerque. None of the Ellis Homestead landowners lived in the canyon year-round (the road is not plowed during the winter, making their property inaccessible except on snowmobile or skis) but were vociferous in demanding dual access to this private inholding, from both the Crest Highway (the quickest route to the canyon from Albuquerque) and Placitas (the quickest route from I-25 and Santa Fe). When the work group representatives decided to ask the New Mexico congressional delegation to intervene on behalf of those of us who wanted to close the road, Congressman Bill Richardson came out in support of our position, while Senator Pete Domenici took the opposite side by supporting the Forest Service proposal to pave the road. (Domenici would later help get Senate Appropriations Committee approval of $600,000 for Phase I of the canyon development plan, the rehabilitation of the picnic grounds and Sandia Man Cave. The entire development plan in the canyon was estimated to cost somewhere between $3 and $5 million.) John Daley was on record as a significant financial contributor to Senator Domenici's reelection campaigns. It was easy to suppose that Paloma Park Development Company's interest in paving the Las Huertas Canyon road was the same as Bob Cooper's, and that the Forest Service decision to pave kept open the door to future ski area development in the second planning period.

The Forest Service often claimed other multiple-use benefits to rationalize road building and commodity production. A major road to a timber sale was presented as a potential cross-country ski trail. Increased grazing allotments were defended because ranchers were required to build more water tanks that benefit wildlife. Clearcuts become wildlife feeding grounds. Though certain aspects of this reasoning may be true, paving Las Huertas Canyon road would essentially encourage increased vehicular use, cause more stream-side degradation, wilderness intrusion, pollution, vandalism, traffic congestion, and the likelihood of commercial development. The only Forest Service answer to these problems was the promise of more personnel in the canyon to enforce parking restrictions and prevent off-road intrusions. There was no guarantee the budget would provide for such services.

At the public hearings on the DEIS, opposition to road improvement and paving was overwhelming. Of the more than 300 people who attended three public meetings, only three gave statements supporting paving (two of these people were Bob Cooper and John Daley). In the Public Comments and Forest Service Response to the Las Huertas DEIS, the vast majority of letters expressed opposition to paving. Many public agencies and elected representatives—including the State Highway Department, local fire and rescue groups, the Sandoval County Commission and Sheriff's Office, and U.S. Representative Bill Richardson—expressed opposition to the proposal.

The final Environmental Impact Statement (EIS) was released in June of 1989. The Forest Service decision was to pave the upper three miles of the canyon road and gravel the lower portion next to the creek. Promoted as a "compromise" decision, this plan was regarded as pre-paving, and sedimentation of Las Huertas Creek would continue to be a problem. The Forest Service maintained that its first obligation was to provide Albuquerque with the highest level of motorized opportunities and to provide access to private inholdings.

Much of the new language in the amended plan addressed the issue of public involvement in Forest Service management decisions. During the negotiations on the Cibola plan, the agency readily admitted there had been a lack of sensitivity on its part to public concerns, particularly with regard to Hispano land grant heirs, who comprise many forest communities, and Native Americans, who use the canyon for religious purposes. Apparently the Forest Service felt that as long as it solicited public comment, it was fulfilling NEPA requirement.

As a member of the work group the Pueblo of Sandia was a staunch ally in our majority decision that the only way to protect the canyon and its resources was with some kind of road closure to restrict the flow of through traffic. The people of the Pueblo have longed used the area to gather materials for use in their religious ceremonies and recognize the canyon's prehistoric significance and riparian importance (the Pueblo was to underwrite the ensuing appeal of the Forest Service decision to develop the canyon). The issue was not so straightforward for the members of San Antonio de Las

Huertas Land Grant. The grant at first viewed any kind of road closure as a further taking of their rights by the Forest Service, after suffering the initial loss of their communal lands when Cibola National Forest became a reality in 1908. While the other members of the Las Huertas work group favored a road closure extending from Placitas to the picnic ground as the best alternative to save the creek and stop the loop flow of traffic through the canyon, grant members didn't want to give up their rights to drive into the canyon to picnic, fish, or pick chokecherries. The grant representative on the committee, Tony Lucero, found himself in an untenable position, unwilling to support closure but also unwilling to be allied with the Forest Service plan to reconstruct and pave the road. The grant took the official position that the canyon should be left as is, with no further development of facilities or reconstruction of the road.

After the release of the final EIS, however, some members of land grant came forward to express their anger at the Forest Service decision and their solidarity with those of us still fighting to have the decision rescinded. Once they realized that the Forest Service would never agree to a "no action" alternative that would leave the road as it was, more members became resigned to the fact that closure was perhaps the only way to save Las Huertas and minimize impacts to the village. They attended several hostile meetings we had with the Forest Service to express their resentment at once again being ignored by the powers that be and attended various rallies and demonstrations called to protest the Forest Service decision.

The Las Huertas EIS was appealed by many of the same groups that were appellants of the Cibola Plan and that were represented on the Las Huertas work group. The appeal stated that the decision was inconsistent with the Cibola Forest Plan, failed to protect Las Huertas Creek, ignored and misrepresented public comment, failed to assess cumulative impacts on environmental resources, and impinged upon the rights of Native Americans to freely practice their religion. In November of 1989 we were notified that all our appeal points had been rejected by the Southwest Regional Forest reviewing officer. The only change in the EIS was to delete a winter closure that was included, as the Forest Service intended to keep the road plowed and maintained all year long (it had never been completely plowed in previous winters). This was the result of an appeal by the canyon landowners who contested even a wintertime partial road closure. Our appeal went to the Chief of the Forest Service, who chose not to review it.

At an emotional protest in Las Huertas Picnic Ground in July of 1991, Steve Sugarman, then an attorney with the Environmental Law Center (owned by Grove Burnett, one of the most successful environmental attorneys in New Mexico), announced that they had filed suit in U.S. District Court to stop the Forest Service from going ahead with its plan to partially pave the road in Las Huertas Canyon. Central to the lawsuit was the contention that the Forest Service had not studied the canyon's resources enough to know what impact developing it would have, a direct violation of NEPA. "If the Forest Service decision has an adverse impact there is no getting the canyon back," Sugarman said in his announcement. Sandia Pueblo religious leader Felipe Lauriano said he *knew* the impact the Forest Service decision would have—centuries of tradition would be destroyed.

As it turned out, the Forest Service's failure to fully consider the canyon's importance to the Pueblo of Sandia as a cultural resource won the lawsuit. After an initial denial of the suit in 1993, the parties to the lawsuit learned that the Forest Service had withheld information (submitted by a Pueblo religious leader and an anthropologist working with the pueblo) from the State Historic Preservation Office about Sandia Pueblo's traditional cultural uses of the canyon that might make the canyon eligible for the National Register of Historic Places. Armed with this new information, the lawyers appeared before the 10th U.S Circuit Court of Appeals in Denver on May 19, 1994 to appeal the dismissal of the lawsuit. In March of 1995, the federal appeals court reversed the dismissal and sent the suit back to the lower court. The parties to the suit immediately sent a letter to the Forest Service demanding a moratorium on any further canyon development and help with the $40,000 in legal fees they had incurred. In October of 1995, the Forest Service finally agreed to hire an ethnographer to survey the canyon to ascertain its eligibility for the National Register of Historic Places. It also agreed to pay the Burnett law firm $50,000 in legal fees.

I used to work in Las Huertas Canyon as a Forest Service fire patrol, which basically meant being a canyon cop: monitoring and directing traffic; trying to control litter; enforcing fire and recreation regulations; building fence and barriers. I was witness to all the problems the work group recognized in the canyon. On heavy use days Las Huertas became a bottleneck of cars, with one lane traffic often the case. Vandalism, littering, theft (in cars parked at Sandia Man Cave), and noise (ghetto blasters in the picnic ground and meadow) were constant problems. One July 4th a fatality occurred just below the picnic ground when a family crashed their car into one of the stone bridges next to the creek (the driver was drunk). Because of traffic congestion in the canyon, both the Placitas rescue squad and I were delayed in reaching the accident. After we'd called for an ambulance (we had to send a driver to Placitas because there is no two-way radio contact in the canyon), it took two hours for it to arrive. That same summer, two other Forest Service employees and I were physically threatened when we attempted to enforce a fire restriction in the meadow.

By the 1980s and '90s I almost never used the canyon. Mark and I never picnicked there, due to the drunks and hotrodders who hogged the road. When I attempted to cross-country ski on the supposedly closed road, I ran into 18-wheelers wrapped around trees, station wagons from Texas stuck in the middle of the road, and four-wheel drive tire tracks as deep as ditches. Occasionally, though, when I used the canyon road for an early morning trip to access a hike on the other side of the mountain, or during a snowstorm while the faint of heart remained locked inside, I had it to myself and saw what historically it has been, to all who have known it intimately.

(Las Huertas Canyon is not on the National Register of Historic Places and the road is still dirt.)

A Last Stand

"What are the roots that clutch, what branches grow out of this stony rubbish? Son of man you cannot say, or guess, for you know only a heap of broken images...."

—T.S. Eliot

The hackneyed expression "tricultural" is often used by those exploiting New Mexico for its diversity, with homogenization in mind. The fact that many cultures, not just three, live side by side but not necessarily together—or equally—is used as a marketing tool to attract both a voyeuristic tourist industry and a group of upper middle class residents whose connection here extends as far as modems and jet airplanes allow a life line out. By the late 1980s, as I found myself in the midst of Forest Service and development battles, I realized that my own little Tunnel Springs community was also changing fast. Suddenly, Mark and I felt like we no longer belonged.

When we first started building in the area just west of the Tunnel Springs road, the only other people out there were Dave and Darri Harrison, who lived next door in a dome. Dave had worked off and on over the years at Zomeworks, with Steve Baer, and had a patent on the bead wall design, which comprised the south wall of their dome: Styrofoam beads, lying between two layers of glass and hooked up to a vacuum cleaner motor, filled up the wall at night or on cloudy days, and receded into a storage space to let the sun in to warm the columns of water used as room dividers. South of us, against the forest boundary, Marsha Latham and Bob Murray built a small adobe house with a windmill to pump their water tank; their neighbors, Christine and Dave Boyd, built a house with solar panels for electricity; to the west Steve Hay ran a wood-working business out of the shop next to his adobe house; to the north, Candy and Bill Frizzell lived in their garage until they and Candy's parents build their modest home complete with chickens and potbellied pigs. While all of us were not necessarily good friends or particularly intimate, we were good neighbors—we watched each other's kids, we loaned tools, we grudgingly chipped in money to periodically get our private road repaired when enough people's car alignments got thrown out of whack in the potholes.

Our first real scare came when Tom and Joanne Ashe expressed interest in buying some land along the road for their own residence. They never did, but it made us all realize how vulnerable we were, with several large tracts of land still available that could be subdivided by the growing number of local developers. We'd been lucky that so far, everyone in the neighborhood had bought small pieces of land for private homes, but as land prices in Placitas began to escalate it became more likely that someone would be interested in buying up the remaining Tunnel Springs land as investment. The other obvious result of increased land prices was the up scaling of the neighborhood, as only those folks with money—professionals, commuters to Albuquerque—could afford to buy our land. A psychologist bought ten acres to the south and built a huge home and garage; to the north, a contractor built a two-story, territorial design house for a young professional couple who worked in Albuquerque; to the east, several houses were built belonging to people I barely knew, just saw coming and going at eight or five. A doctor and his wife did build an Earthship house, which complimented the eclectic array of original houses, but ended up suing their contractor when all their windows began to crack.

And what all these people wanted—people with mortgages, people hurrying back and forth to work—were more services and amenities. The disrepair of the road, which the rest of us valued as a deterrent to unwanted traffic, became a huge issue of contention as the newcomers lobbied to have it upgraded, graveled, or even paved. They wanted a road maintenance agreement to make it easier to get mortgages. They wanted police protection. They wanted cable TV. Mark and I began to feel more and more estranged, not only in our Tunnel Springs community, but in the larger Placitas community as well, where only a few neighborhoods like Dome Valley or the village itself retained any funky soul. Many of our original Placitas friends had already left; those of us who stayed struggled to define and defend what we thought we were in an increasingly complex situation.

When Jakob, our first born, was six, we decided to have another child.

When I think back on it now, perhaps one reason we made that decision, when it would have been so much easier to rest with one, was an attempt to recreate the feelings and atmosphere of a time in Placitas when we found fifty people to invite to Jakob's first birthday party—kids and adults—who all liked each other. Later on, that was not so easy, as friendships strained and camps divided between those who fought the subdividers and those who joined their ranks. It got to the point where I had to ask who would be at a party before I was willing to attend. Not an auspicious time to settle in with another child, but then when is it ever.

I suffered a miscarriage in our first attempt, in 1987, but by the beginning of 1988 I was pregnant and nauseous, always a good sign. How I managed to make it through the cross-country ski classes I taught all January and February, I don't know. I guess staying physically active kept my mind off my predicament as I scurried around trying to keep everyone warm and happy while we skied the Sandias, Jemez, and Sangre de Cristo mountains, learning the basics of cross-country technique. My February class, an intermediate level, was always more fun than the January beginner sessions, as we skied some of the more difficult routes, like the Winsor Trail out of the Santa Fe ski basin.

March and April were miserable, as the nausea continued into my fourth month. The only way to settle my stomach was to constantly eat— crackers, carrots, cheese, pretzels, bagels, whatever came my way. I gained weight, started to lose my winter muscle tone, and despaired of ever feeling well again. My midwife, Nancy Middlemiss, was my friend and ally through it all, and her husband, Marcos Sanazaro, a resident at University Hospital, my savior when I developed migraine headaches along with the nausea. I was seeing Nancy for my prenatal care, although we both knew I would have to have a hospital delivery, as was the law in New Mexico then, when attempting a VBAC—vaginal birth after Caesarean. Jakob had been born by Caesarean section after a very long and depressing labor that left me only three centimeters dilated and a baby "sky high," as the doctor described it (I was supposed to have Jakob at Southwest Maternity Center, but ended up at Presbyterian Hospital, upset and pissed off). Nancy would see me through to the last few weeks, when a sympathetic doctor, Rebecca Jackson, would take over for the hopefully low-key, successful delivery.

Fortunately, by June the nausea subsided and I was out on the trail, leading my hiking groups for the University of New Mexico's continuing education program. I'd been leading hikes in the Sandia Mountains for eleven years, a wonderful way to earn money as well as stay in shape. People joined my hiking groups for a variety of reasons—to learn more about the Sandias, for exercise, to study the wildflowers with me, to socialize. I've stayed in touch with several of them over the years, and even bartered various services for hikes—child care, massages, and a hernia operation for Mark (that took several years of hiking and skiing classes). Mark, Jakob, and I went on a backpack trip that summer, up the East Fork Trail from Red River to Lost Lake. It was an interesting experience to trade places with Jakob as the trip whiner, as he hiked out in front, periodically turning back to check my progress and convey his encouragement. On a camping trip in August, after two nights of no sleep on the hard ground, I decided I'd had enough adventure for the year, and confined my physical activity in September and October to morning walks with Jakob to catch the school bus.

My water broke about nine o'clock in the evening on the 17th of October. Because I was trying for a VBAC, Becky, my doctor, wanted me to come to the hospital that night so I could be closely monitored. My contractions started in the car on the way into Albuquerque, which made me euphoric, as I'd never even gone into labor on my own with Jakob. I thought for sure they portended a normal birth. I was checked into a room, and after Becky took a look (almost no dilation), both she and Mark went to sleep while I walked around the room, squatted, lay down, did whatever I could to pass the time until the contractions became strong enough to give me pause. I spent most of the next day in bed, as the contractions came fast and furiously. Becky stayed with me and gave encouragement, while Mark and I tried to breathe. As I had learned with Jakob, most of the breathing techniques taught in our Lamaze class really didn't do that much to keep me focused, and I developed my own breathing routines to try to keep from screaming. Reluctant to check the progress of my dilation too often (it hurt horribly), Becky remained hopeful that soon I would start to dilate as the contractions became more intense. Periodically, the nurses hooked me up to a fetal monitor to check the baby's heart, which remained strong. In the next room I could hear a steady moaning as a voice said repeatedly, "Don't push."

I was in labor for fifteen hours, and while I knew I wasn't going to die, by about two o'clock in the afternoon I wanted to. After fifteen hours and only three centimeters dilation, I was mad at everyone, and afraid. The contractions lasted for over two minutes, with only thirty seconds rest in between. I yelled at Mark when he tried to make me breathe, and I cursed Becky when she told me to try and imagine I was somewhere beautiful and calm, like on a beach, lying in the waves, watching the palm trees. I told her I didn't care where I was, I just wanted the fuck out of there. She looked at Mark and told him, "We're usually pretty close when they lose their sense of humor."

What we were close to, though, was the emergency mode. The fetal monitor was on me constantly now, and as Mark watched the pattern of my contractions on a screen ("That one wasn't as bad as the last one," he said, as I reared up off the bed in agony), suddenly the baby's heart rate dropped and Becky said, "That's not good." The next contraction showed the heart rate back up, and everyone relaxed. Then several contractions later, down it went again, but after another contraction, came back up. Several more nurses came in, and Becky quickly sent one of them from the room. She came back several minutes later with the resident obstetrician, a beautiful young Hispana. She introduced herself to me and conferred with Becky. She decided to insert a vaginal monitor that could more closely calculate the baby's heart rate, and I screamed as she inserted it. My dilation was still only three centimeters, and she told me that she wanted to put me on a drug called Pitocin to see if they couldn't get things moving more quickly. I knew what Pitocin was, having had it with Jakob, and I told her I didn't think I could stand the intensity of the contractions it caused (as if anything could be worse that what I was already experiencing). She decided to give me an epidural, a sedative that numbs you from the hips down, usually administered in preparation for a Caesarean. Then, just as she was leaving the room to order the epidural, the heart rate dropped again. This time it didn't come back up. One of the nurses remarked, "This is not a happy camper."

Suddenly the doctor said, "That's it," and they had me on a stretcher running down the hall to the operating room, an oxygen mask slapped on my face. There, in a sterile room filled to the brim with people, they shaved me, poked me, prodded me, stripped me of any modesty that might have managed to linger that far. Mark was there beside me, and when they told me to count to four and I'd be out, I silently said goodbye to him, just in case.

Jakob and Max.

Mark told me much later that he stayed in the operating room while they cut me open, but when they couldn't get the baby out of the incision (they reused the one Jakob came out of), and two people had to pull and two people had to push, he left the room, crying. A few minutes later Becky went out to him with the baby in her arms, alive and all there. Forty-five minutes later I woke up in a small, white room, and they brought me Max, suctioned, cleaned of my vernix and fluids; I'd missed the whole thing, my body a piece of meat on the table. But there he was, crying and mourning this shocking, painful taste of life outside the womb, and he immediately attached to my breast, desperate for comfort. He didn't sleep for the next ten months and developed a personality to match—always at attention. He seemed to match the tenor of the times, as what would be our last few years in Placitas became filled with much the same intensity of being.

The Pueblo of Sandia

"The Mountain is our church, our protector, and our source of
well-being. We believe that many of our blessings derive from the
Mountain, and it hurts us when we see it abused."
—Joe M. Lujan, Pueblo of Sandia Governor

The Sandia Mountains define the eastern horizon above the sprawling
metropolis of Albuquerque. Civilization creeps up the mountain
foothills as expensive homes get built on every available hill and
arroyo, invading critical wildlife habitat, including that of the Rocky
Mountain bighorn sheep. The city and Cibola National Forest have acquired
only small segments of open space land where the wild and free still abide. In
this mess of development, thank goodness for the Pueblo of Sandia. Nestled
between the Rio Grande to the west and the Sandia Mountains to the east,
this large Indian pueblo (24,187 acres) halts the urban sprawl from the
boundary of Albuquerque to the north end of the Sandia Mountains—the
only foothills land left to the wide-open space.

The Sandias have long figured in Pueblo cosmology. The mountains
contain sacred shrines and areas used for the collection of resources necessary
to the Pueblo's traditional religion and culture, and have provided water,
shelter, and refuge, as well as game and wild plants for food, medicine, and
ceremonial use. Because of its special relationship with the Sandias, the
Pueblo has always been an ally—spiritually as well as financially—to those
who have fought to protect these mountains from various threats: more
housing developments, increased traffic and congestion, and the potential
destruction of the unique riparian area in Las Huertas Canyon. As I discussed
in previous chapters, the Pueblo was party to the Cibola Forest Plan appeal,
underwrote the appeal of the Las Huertas Canyon Environmental Impact
Statement, and was one of six parties to the Las Huertas Canyon lawsuit.

Yet, in 1986, when the Pueblo attempted to rectify an inaccurate
1859 survey that deleted 9,480 acres of the Sandia Mountains from its

original Spanish land grant, the Pueblo found an unlikely opponent: the environmental community. As land grant historian Malcolm Ebright explains in his book *Four Square Leagues, Pueblo Indian Land in New Mexico*, in 1748 the Spanish governor ordered his general to go to the pueblo and lay out its claim in the traditional four square league, meaning a square of land, each of whose sides is one Spanish league distant from the center of the village (a league measures approximately 2.6 miles). Because the natural western boundary of the Rio Grande lay closer to the pueblo than a Spanish league, the northern and southern boundaries were extended beyond a league to make up the difference. The eastern boundary, the Sandias, was thus described: "and on the east, the Sierra Madre, the Sandia, with which boundaries are the conveniences of pastures, woods and watering places . . . in order to maintain their stock, both large and small, and a horse herd."

The Pueblo also contended that the subsequent 1864 survey patent incorrectly described the eastern boundary of the Pueblo included in the 1748 Spanish Land Grant. Because the Pueblo's claim included Forest Service land that extended from a Sandia Mountain sub range to the very crest of the mountains, the Pueblo found itself an adversary of groups like The Wildlife Federation as well as local conservation groups. Conversely, by opposing the Pueblo's land claim, environmentalists found themselves in collusion with not only the Forest Service, the agency whose policies they had been fighting for years, but wealthy landowners whose subdivision homes were part of the claimed acreage.

All too often this is the scenario in Indian land claims. Many years ago some of these same groups opposed one of the hardest fought Indian battles waged by Taos Pueblo for its sacred Blue Lake in northern New Mexico. When appropriated by the Forest Service as part of Carson National Forest in 1906, the Pueblo lost its most sacred shrine, site of numerous religious rituals. Although the Forest Service allowed the Indians special use of the lake, the Pueblo fought persistently for its reacquisition, intensifying its efforts in the 1960s because of increased timbering and recreational demands. In 1970, Congress passed a bill recognizing the Pueblo's claim to the lake and 48,000 surrounding acres. Since then, the area has been zealously protected by Taos Pueblo, off-limits to the general public and even those from the Pueblo not specifically sanctioned.

This off-limits status, more than anything, is what raised the ire of conservation groups that opposed the Blue Lake claim in 1970 and those groups in opposition to the Pueblo of Sandia claim. Viola Miller, then chairwoman of the Albuquerque Open Space Task Force, expressed the environmentalists' sentiments against the Pueblo when she said, "If we lose this Forest Service land and this wilderness area land, we basically will not have control of its future use for generations and years to come."[23] Clifford Mendel, Albuquerque Wildlife Federation president, added that "the transfer will spell the development of the nineteen percent of the Sandia Wilderness claimed by the pueblo."[24] These are the same voices that continually claimed, in appeals and lawsuits, that the Forest Service wasn't being responsive to public input regarding management of forest lands and had itself capitulated to development interests.

One of the most active groups during the late eighties was called the Sandia Mountain Coalition, comprised primarily of landowners in the Sandia Heights, Tierra Monte, and Evergreen Hills subdivisions. The group raised over $10,000 by collecting $200 from the families living in the subdivisions to retain an attorney, hire an archeologist (to study some disputed boundary markers), a surveyor, and historian (to study the original Spanish and U.S. documents outlining the Pueblo boundaries), as well as underwrite a trip to Washington, DC by coalition members to argue their opposition to the claim before the Secretary of the Interior, which they did, in July of 1988.

As early as March of 1988 Joe. M. Lujan, then governor of the Pueblo of Sandia, issued a letter to the landowners stating that the Pueblo would recognize "all existing legitimate rights on the west face of the Mountain as to all privately-owned lands, rights-of-way, and other property interests."[25] Rights of ways for various utilities would also be recognized, and no property taxes or zoning restrictions would be imposed. There would be no change with respect to criminal jurisdiction on private lands as well. The Pueblo again reiterated its desire to keep the forests lands in a "natural, unspoiled state in keeping with the religion and tradition of the Pueblo. The Pueblo is developing, in conjunction with the Department of the Interior, a comprehensive management plan which will regulate the use of the area. The plan would provide for regulated public use."[26] The Pueblo also assured the landowners that if its claim were recognized, it would negotiate with the private landowners to retain title to their land holdings.

· During these months of lobbying and rallying against the Pueblo's claim, somehow the rumor started that I was employed by the Pueblo of Sandia and was promoting the land claim. I got telephone calls from several outraged members of a group called Save Our Sandias, who insisted that all the Indians wanted was to "get something for nothing" and deny the public access to the Sandias. They were tired of Indians trying to capitalize upon white guilt to get whatever they wanted so they could use it for their own selfish purposes. I spent an hour on the phone with a spokesperson for the Sandia Mountain Coalition (who also called thinking I was a Pueblo employee) explaining that while I wasn't employed by the Pueblo, I certainly supported its land claim and thought all environmentalists should as well. I perhaps convinced her that I was speaking for myself, but I certainly didn't impress upon her that the enemy was the Forest Service, not the Pueblo. I did get one sympathetic call from an old friend, Dennis Feeney, a University of New Mexico professor who was then living in Sandia Heights. He told me he'd had much better luck with his Pueblo neighbor than with his other neighbor, the Forest Service, and fully supported the Pueblo claim.

The Forest Service, of course, immediately denied merit of the Pueblo's claim. In a reply to the Pueblo in June of 1987, it contended that the 1864 survey patent correctly identifies the boundaries and supported that contention by arguing that the amount of land included in the survey, 24,187 acres, was already larger than a "standard" Pueblo grant. The Forest Service went on to say that because some of the lands the Pueblo claimed were federal lands, Indian title, if any, has already been extinguished. And any transfer of lands to the Pueblo "could result in a loss of public access rights" and "could cause private land owners to lose their land."

It seems logical that environmental groups would actively support Native American land claims as the ultimate means of maintaining forest resources, with or without public access. Taos Pueblo has maintained Blue Lake and its surrounding forest lands in as pristine condition as we will ever know them. The Pueblo of Sandia stated in a position paper concerning its land claim: "The Sandia Mountain is sacred to the people of the Pueblo of Sandia. The Mountain is our church, our protector, and our source of well-being. The integrity of our community and our political system and the maintenance and transmission of our cultural identity depend on that relationship." One would think, with their emphasis on wilderness preservation, that the environmental groups would gladly accept more restricted access to these mountain areas because the Pueblo values the mountains—its waters, wildlife, trees, and wildness—in a way the Forest Service will never comprehend, much less manage. And in a political sense, many feel any aboriginal land claim or Hispano land grant claim, be it 9,000 acres or 50,000, should be recognized as compensation for the unjustifiable abrogation of this country's treaties with Indian peoples and the criminal treatment they have suffered as second-class citizens.

After the signing of Taos Pueblo's precedent setting claim, the federal government seemed ready to assume the role of Indian advocate with the passage of Public Law 93-638, the self-determination initiative that opened the door for millions of acres to be returned to Indians. The Pueblo of Sandia's land claim, however, languished in the labyrinth of the Department of the Interior for five years. In 1988, Ross Swimmer, then Assistant Secretary for Indian Affairs of the Department of the Interior, was quoted as saying that he believed the Pueblo had a legitimate claim to the west face of the Sandia Mountains. However, a decision against the Pueblo was rendered by Interior in 1989, and in the winter of 1994, Interior Secretary Bruce Babbitt refused to overturn that opinion and the time limit for addressing the claim within the Department of the Interior expired. The only recourse left was litigation, and in December of 1994, the Pueblo filed suit in federal court in Washington, DC.

The Pueblo had been reluctant to involve New Mexico's congressional delegation in the Department of Interior's review (according to Babbitt, Senator Domenici advised him not to overturn the opinion). As tribal administrator Malcolm Montoya pointed out, "Our claim, unlike that of the Taos Pueblo people, affects a mountain range that is adjacent to a huge population that uses that land extensively. It's not a popular issue."[27] Despite opposition from the congressional delegation, the Pueblo of Sandia continued to state its intent: "Our commitment is to keep the mountain land in a natural, unspoiled state. The intent of the federal government seems to be the opposite."[28]

In 1998 a federal court decision validated the Pueblo's land claim to the Sandias. The Forest Service immediately appealed the decision at the behest of the state's congressional delegation. In a 2000 negotiated settlement between the Pueblo, Forest Service, and Sandia Peak Tramway the Pueblo agreed to relinquish any claim of ownership in return for a perpetual right of usage for traditional and cultural purposes, as well as the right to approve or withhold consent to new uses and the right to be consulted on modifications to existing uses. Again, the congressional delegation objected to the decision and declared they would fight its passage through Congress. Finally, in 2002, after the Pueblo agreed to purchase a 160-acre track of land in the middle of the wilderness that was owned by Albuquerque businessmen, a settlement agreement was signed by all parties and Congress passed a law enacting the agreement.

We often went to the dances at the Pueblo, at harvest time or Christmas time, always open to anyone who wanted to share in the celebration of the occasion. The Pueblo sits in the midst of fields by the Rio Grande. It is home to only several hundred inhabitants, and it retains a feeling of antiquity, of what has been before, of what cannot be destroyed. As in the other pueblos scattered along the river, whose people share a common heritage, the dancers, divided by clan, fill the plaza during the dance periods. Lines of dancers snake in and out down the length of the plaza to the steady beat created by a circle of drummers and singers. Their choreographed movement is somehow both solemn and joyous. Beneath the shadow of the mountains, it's always an impressive and moving sight. People watch quietly for hours, before moving off to wander down the row of booths with Indian arts or homemade *tamales* or *burritos*.

Because of our political involvement with the Pueblo someone we had worked with would invite us into a home to share a meal. Anyone, really, can enter a home and eat there, because that is the way things are done at the pueblos, sharing both their dances and their food with anyone who shows up. You always feel a little shy at first as you take your place at the table, where food is constantly consumed and replaced from the nearby kitchen by an army of Pueblo women who've been cooking all day and will probably be cooking all night. But everyone is eating and laughing and visiting, and you join in with gusto. Red and green *chile* in stews, on potatoes, in *burritos* or *enchiladas* are sopped up with oven bread and *tortillas*, while green salads and fruit salads and ears of corn and bowls of green beans provide cool relief.

Pies and cakes and cookies and custards satisfy the sweet tooth, and it's all washed down with lemonade or punch or coffee or tea before you stagger back outside to watch the dancers in the bright sunlight. Sweat is pouring off their faces as they continue their mesmerizing dance through the long hours of the afternoon. This is the way they show their commitment to the mountains.

Exodus

"I came to Placitas to live the kind of life already *here*. Those folks who have been moving here recently seem to want the kind of life they came *from*."

—Tom Nordstrom

One day in 1989 the wife of one of the developers called up to castigate Mark and me for writing a letter to the *Albuquerque Journal* in response to a column Slim Randles did on all the billboards around town promoting the Placitas Land Co. In the article, the developers promoted themselves as upstanding members of the community "who aren't getting rich."[29] The woman called to tell me that if we'd just keep quiet everyone in the community would be happy and content.

We weren't the only one writing letters, though. A group calling itself "Citizens for Rural Placitas" wrote the following letter to the *Albuquerque Tribune* in November of 1988:[30]

"We are a group of Placitas community residents who have chosen to call ourselves Citizens for Rural Placitas. We represent those members of the community who must speak out against the short-sighted exploitation of Placitas and its environs by land developers and entrepreneurs. Articles, billboards, and extensive advertisements have been generated by real estate opportunists, spotlighting Placitas with the clear intention of aggressively encouraging property sales in the area. Citizens for Rural Placitas deplore the callous disregard shown by these developers of the needs and concerns of native and long-time residents of the community.

"Even with Albuquerque's growth and an increasing market for real estate in communities within commuting distance to the city, realtors, developers, and builders should respect the integrity of *all* those residents of Placitas who think of the village as a unique community, a hometown with character, a place of diverse cultures and lifestyles, and a peaceful and quiet place to live.

"In their advertisements, developers have been stressing the traditional and the historic aspects of Placitas to lure prospective buyers, not caring that the increased traffic, noise, crime, will destroy those very qualities we as residents strive to preserve in our village. Hispano residents and land grant heirs are being exploited as part of the village's 'quaintness' while their ancestral land is being sold at prices only the privileged can afford.

"The realtors also advertise that 'undesirable' alternative lifestyle groups have moved on, while at the same time exploiting the image of Placitas as a community of artists and crafts people, many of whom are the same individualists the realtors say (and hope) are gone. We find this hypocrisy repugnant.

"Early development of Placitas was characterized by individuals who conceived their homes as places they would live in for the rest of their lives. These homes were often built over a period of years, each one growing and changing to fit the needs of a particular family. The covenants of the new developments discourage farming and traditional lifestyle. The idea of house and land as marketplace commodity contrasts sharply with older values and the more traditional style of rural life. We Citizens for Rural Placitas feel that most long-time residents value land for the beauty and sustenance it provides, not for the profit their investment might turn tomorrow.

"We feel it is important for the members of the Placitas community to understand that the real estate developers do not have the best interest of the community at heart, no matter what they may say to the contrary. We do not want Placitas to become just another over populated development, its character and soul gutted by developers. But time is running out. The power of the buck now rules those who see every acre of undeveloped land as another rung on their climb to personal riches. Will we be left with a ruined Placitas, while those who now advertise our community with such fanaticism leave for greener pastures and another rural setting to exploit?"

This letter was indeed prophetic. By the time it appeared, some of the developers were already on their way to "greener pastures." Pepi and Rick Levin no longer lived in Placitas, Steve and Wendy Gudelj would soon move to Albuquerque, the Ashes bought a second home in Martha's Vineyard, and Willy Levin, Rick's brother, who had moved to Placitas to be a contractor in the subdevelopments, moved back to Seattle.

By 1989 Mark and I were talking about leaving, too. Tired of constant battles and scared by the fact that neighbors' wells were going dry, we started looking northward, to more rural communities, where we could buy some irrigated land and hopefully live the life we had originally moved to Placitas to find. Because we had started a small publishing company and both wrote freelance, we had the luxury of relocating to a more remote part of New Mexico where commuting to work in town was not an option. We were ambivalent about this decision; reluctant to leave our home, literally our sweat and blood, and apprehensive about moving into a new community and being the outsiders ourselves. Although we felt we had paid our dues in Placitas, wherever we ended up we knew there would naturally be a certain amount of suspicion and distrust directed our way.

As we talked more seriously of leaving, I was comforted by the thought that Scott and Helen Nearing, long-time organic farmers and socialists, had, in their seventies, left their home in New England when a ski area moved in and started *another* home in a more remote seaside village where they could live in peace. Many of our friends had already moved on, some in search, like us, of a more remote area to try to sustain an alternative way of life. Others remained but began pursuing more conventional careers as their families grew and it became harder to live as marginally as many of us had managed throughout the 70s and 80s.

After a while, there's not much left to fight for. In place of community is anonymity. We didn't know many of our neighbors, much less the people in the subdivisions. When Placitas was smaller you could choose not to be more than an acquaintance to the neighbors next door, but you certainly knew who they were and what they were doing. And when you know who your neighbors are, your actions become more responsible and your values and theirs are not compromised.

The final straw personally, I think, was seeing our new neighbor out on his land digging holes for a chain-link fence to surround his 2.5 acres. A California transplant who worked for the computer industry, this neighbor brought with him a borrowed rationale "fences make better neighbors" that may have been applicable in California, but not Tunnel Springs. When his intentions became clear, the entire neighborhood wrote a letter asking him to please reconsider his fence—not only was it out of place in an area of wide open space (where an occasional adobe wall or coyote fence demarcated a minuscule patch of grass), it was unnecessary. We, as a neighborhood, would watch out for suspicious characters intent upon ripping-off his garage (which he intended to build before his house, to store his tools and supplies), a fear he expressed in response to our letter. When he refused to reconsider the cyclone fence, we upped the ante with a letter from a neighborhood lawyer implying that we thought the fence might be in violation of covenants on the land that restricted "structures not in keeping with a Southwestern style" (whatever that means). This particular piece of land actually had some covenants attached to it because it had been sold several times previously and somewhere down the line one of the sellers had written restrictions into the deed. When our neighbor received the lawyer's letter, he threatened to sue some of the other people whose land had covenants for noncompliance of various restrictions. At this point, reluctant to spend hundreds of dollars to hire the lawyer to see if we could prove the man was not in compliance with his covenants, everyone threw up their hands in disgust and retreated. The fence went up in due time, and I guess if this neighbor's idea of "good neighbors" were those who didn't speak to you, he has been vindicated.

When we finally decided to sell our land in earnest, we divided our property into two parcels, one with the house and one as vacant land. Many of the people who came to look at the land lived in the subdivisions. One older couple declared, "I want to live someplace where I never have to go to another Neighborhood Association Meeting as long as I live." Another younger couple told us, "We didn't find out that the house we bought in one of the subdivisions wasn't an adobe until we moved in." Naive they obviously were, taken in by subdivision lingo that promotes "adobe style" or "solar adobe" houses if one adobe trombe wall is in place, but it's still false advertising. Another couple told us, "One acre seems like a lot of land until a builder starts a new house right out your bedroom window, and there's nothing between you and your neighbor except a lot of sand and a terrific view of your bed."

We sold the land to an upwardly mobile computer couple who were into high-tech recreation. We sold our house to a family that owned a law firm in Albuquerque but had Sandoval County political aspirations. We moved out at the end of the school year, 1992. Several years after we moved we learned that the people who bought our house removed the entire second story, built a 2,000 square foot addition, and then reroofed the entire thing. One of my neighbors told me this when I ran into him down in Bernalillo one day: "You can't even see that your house is in there," he said, as my broken heart sunk down into a wrenched gut.

Norteño Life

"Why Taos? Why northern New Mexico? I mean, every time I enter my criminally potholed driveway behind the wheel of my decrepit VW bus, the odds are seven to five I won't even make it as far as the corrugated culvert that channels the waters of the Pacheco ditch under my driveway and into Tom Trujillo's property.... Yet I have a quirk inside. I kind of like it when the natural world clobbers our technological universe. So what if hurricanes, earthquakes, mosquitoes, and idiot cattle make my life miserable? Thank God nature can still kick me and all my gadgets in the teeth!"

—John Nichols

My neighbor's tractor lights shine through the bedroom window as he cuts the hay in the back pasture. He begins at eight in the evening and finishes around ten-thirty, after a day's work at his full-time job in a city an hour away. It's an incredibly long day, and he's incredibly tired, but when the hay is ready and there's a succession of rainless days in which to cut and bale it, it's now or never.

In this more rural, more small-town community of El Valle, people like my neighbor struggle to maintain a farming way of life amidst 20th century economic pressures that force local residents to leave home or commute to jobs in town. (There is a long tradition of men leaving these villages to earn a living herding sheep in Wyoming or working in the mines of Colorado.) Along with those who are able to eke out an existence without outside jobs, we settle for the essentials: a ten-year old car that needs a new muffler; a second-hand refrigerator when the first one breaks down; rare trips to town for dinner or a concert.

While there's definitely a sense of protectiveness here, and insularity, everyone has been incredibly warm and generous to us, even when we showed up as an unproven family, brand new to a village whose Hispano families have generations of attachment. They have learned over the years that the outsiders who moved here—and there weren't many—did so because they wanted to be part of small town life, where neighborliness—being a *buen vecino*—is valued.

It's not rural chic here, either. The homes range from the traditional tin-roofed, train-car style adobes painted lovely greens and pinks to plastered-in trailers with wood stoves. The handmade houses of the Anglos, who first settled the communes forty years ago, are a hodgepodge of domes, greenhouses, rambling adobes, and freeform structures of futuristic design. They all sit in fields of timothy grass, grazed by cows and horses, with at least one mandatory junked car on display. It's probably the most beautiful landscape in New Mexico, if not the world: the 13,000-foot peaks of the Sangre de Cristo Mountains provide the backdrop, the piñon-juniper hills rolling to sandstone cliffs the vista. At almost 8,000 feet in elevation the summers stay cool and lush while the winters intimidate those whose blood hasn't run cold for three hundred years.

We immediately felt at home. We learned how to guide the water from the *acequias* to our fields of hay, and that garlic grows great at 8,000 feet. We planted another orchard, built up the soil for another garden, and established the second biggest woodpile in the village. We've been commissioners on one of our *acequias*, and in the summer and fall we help our neighbor bale and load a lot of hay.

But the luxury of trusting those who move here may be short lived. While in Placitas the developers came and conquered, I'm afraid they're already in this backyard. Taos, the closest town of any size, is locked in battles against airport expansion (for 727's winging their way in from New York and California), golf courses, luxury home development—all tourist attractions seen by city fathers as the salvation of a depressed economy. The price is obvious—crowded streets and stressed-out services, farming lands turned residential and recreational, loss of cultural identity.

Will the development overflow to us, forty-five minutes away over a spectacular mountain road? Probably. But it will take a while to inroad this community whose isolation and hard-boiled reputation stalls the suburban spread. Despite the friendliness we have found, there is an undercurrent of *unfriendliness* towards those who drive their fancy cars through town on their tours of northern New Mexico, and those who are ready to set up their modems and beam to L.A.

What is it about these villages of northern New Mexico—San Cristobal, Peñasco, Llano, Chamisal, Las Trampas, Mora, Guadalupita—that so startles the imagination? Life here has been chronicled by John Nichols, Enrique de la Madrid, Stanley Crawford, Sylvia Rodriguez, and William deBuys in inspired fiction, folkloric tales, reverential fact, and anthropologic study; all have been able to portray the sense of community and place without romanticizing or ignoring the very real hardships and limited economic viability of these villages. Mark and I knew we were in a privileged position to be able to move here when so many of the local people have to move away to find employment. But for everyone who leaves (and few of them ever sell their land, for they plan to return someday), those who stay continue to maintain their land and way of life, which define who they are. It's obviously not an option for everyone. How many of us have incomes not dependent upon an eight-to-five job that exists only in an urban area? Yet it is also a question of priorities. Mark and I made the decision to live on the very limited income our self-employment creates; the people of northern New Mexico, who return home with a fierce loyalty, irrigating bean and hay fields, do so not because they can't join the urban dwellers but because it is here that their relationship to the land and to each other—however complicated and fractured that can be at times—is realized.

While all of us are trying to chronicle, and by doing so, defend the vitality of these mountain villages, we should also be talking about what we can do to insure that these communities survive into the twenty-first century. In an article for *High Country News*, Ed Quillen, former columnist for the *Denver Post* and publisher of *Colorado Central* in Salida (Ed died in 2013), talks about a Homelands Commons paradigm in the Rocky Mountain west: "Is there a way to enjoy the virtues of a sustainable culture, while allowing individuals to be individuals?"[31] In other words, can the Hispano land grant heirs, whose families have lived in these villages for 300 years, find ways to sustain an agricultural base, pursue their arts and crafts, and find jobs in a local economy? Can those of us who aspire to a remote, quiet life in the mountains integrate ourselves into these communities while offering our support and creativity? Will the powers that be—Forest Service, land planners, county governments—ever learn to say no to sprawling housing developments and resorts; say no to corporate timber sales that destroy forest resources; say no to more galleries, boutiques, and ski areas that raise land prices, tax people off the land, and create a playground for the rich and famous who have "not a response to a place, but rather to [their] affluence and social status," as Wendell Berry describes them?

Our neighbors Tomás and Fred Montoya cutting our hay.

There is a long history of activism in northern New Mexico, those who have tried to address these issues in a real and positive way. In Los Ojos, near Tierra Amarilla, Ganados del Valle (Livestock Growers of the Valley), a wool cooperative that raised sheep, processed the wool, and wove it into finished products, provided jobs and security for many of the local people in this traditionally depressed area. Established by eight people in 1983 as a non-profit economic development corporation, their first business was Tierra Wools, run by locally trained weavers who spun local wool into the traditional Rio Grande style blankets, rugs, tapestries, and clothing. Tierra Wools sells its products from a retail showroom and workshop in Los Ojos. Sales in 1990 were over $200,000, with twenty-four people employed.

Agricultural support programs with a revolving loan fund enabled Ganados to help the small sheepherders in the area who supplied the wool for the weavers. Two hundred sheep were acquired as part of the *partido*, or livestock shares program. Under this program ten to fifteen sheep were "invested" with a grower for six years. Each year the grower was required to return one lamb to the organization. The lambs in turn were reinvested with another grower or sold to fund a scholarship program. By the end of the sixth year the original borrower was required to return to the organization the same number of sheep originally invested.

In 1988 Ganados used revolving funds of $80,000 to purchase and create the Pastores Feed and General Store. By marketing livestock feeds, veterinary supplies, handmade gifts and foods, monies made from the community circulate one more time through the local economy. Family-based enterprises were encouraged to test the waters at the store with their homemade tortillas, fresh eggs, jam, quilts, and woodcarvings. Another 1988 enterprise, Pastores Lamb, began marketing commercial and specialty lamb to restaurants in Santa Fe and elsewhere. In 1990 Pastores Lamb received certification as an organic lamb grower. Another business, the Rio Arriba Wool Washing Plant, washed over 2,000 pounds of wool that Tierra Wool wove into its products. A market plan was being developed to attract Rocky Mountain and west coast consumers. In 1996, Tierra Wools became independent of Ganados and established itself as a viable for-profit business.

While Ganados del Valle was a showcase for a "regenerative economy"—a self-reliant economy that can renew itself, the organization saw controversy over some of its practices. Part of developing a sustainable economy is retaining local control over land and water uses. Ganados has been an advocate for the integrity of agricultural lands and the *acequia* system, battling to strengthen subdivision regulations and oppose a proposed ski area. A consistent problem for the association has been the shortage of available grazing land to feed the cooperative's herds of sheep used in its various enterprises. Most of the land around Los Ojos and Tierra Amarilla is privately owned, and most of that is rented to cattle growers. The only other available land, 45,000 acres managed as wildlife refuges by the New Mexico Game and Fish Commission, is part of the long-disputed common lands of the Tierra Amarilla Land Grant, made private by Congress in 1860. Beginning in 1982 Ganados del Valle became involved in negotiations for use of this land in a grazing research project. They proposed that they be allowed to graze sheep on the land in a controlled experiment to establish some scientific fact as to whether sheep and elk grazing could be compatible. After extensive negotiation and lobbying, their petition was denied, and in August of 1989, led by Antonio Manzanares, 1,000 ewes and lambs were moved onto the W.H. Humphries Wildlife Management Area.

Antonio Manzanares

While "breaking the law"[32] didn't appeal to Antonio Manzanares, neither did the idea of abandoning a livelihood that had kept him in the Valley where his family had lived for over a hundred years. When he led his sheep onto the wildlife management area, it was an act born of many years frustration—frustration at seeing the young people of the Chama Valley leave because of no opportunity for work, and frustration at the lack of summer pasture for his sheep. Every year, as he struggled to find pasture on national forest land, Jicarilla Apache land, or the large land holdings of Anglo ranchers, he threatened to sell out. The occupation of the wildlife area was both a short term attempt at making it through another summer as well as an attempt to resolve the issue of whether indigenous people in a poor area have a right to use public lands that were formally part of their Spanish land grant. The latter has not been resolved, but Manzanares continues to raise sheep for economic and cultural survival.

The sheep ranchers were immediately cited by officers of the New Mexico Game and Fish Department, but Ganados board members decided that the sheep would stay put until they had somewhere to go. Under the spotlight of the national media, Governor Garrey Carruthers came up with the short-term solution of allowing the sheep to graze on a nearby state park for a few weeks and hastily put together a ten-member task force representing various interests and viewpoints. John Fowler of New Mexico State University was to direct the group to conduct a study of whether livestock could enhance habitat for wildlife. But the group soon disintegrated into discord, with hard-core environmentalists and state agencies adamant about not letting sheep on the wildlife lands so the research could be done, and Fowler's study never materialized.

Maria Varela

Maria came to New Mexico in 1967 to work with land grant activist Reies López Tijerina, straight from years in the South with the Student Nonviolent Coordinating Committee. Politically experienced and sophisticated, she saw land tenure and water rights issues as the core of poor people's struggle to retain their cultural identity and livelihood. Making a home for herself in the Chama Valley, Varela helped start La Cooperative Angola and became director of La Clinica before joining with Manzanares to organize Ganados del Valle. According to Manzanares, "Maria was the visionary," while he was the "pessimist, just trying to make a living."[33] The combination worked, however, and although Varela expressed anger and frustration at the environmentalists who opposed Ganados on the public lands task force—"we ought to be together in fighting development, which is the real enemy of wildlife"[34]—she remained resolute in her search for land that would allow for further economic growth for the people of Ganados.

Once again environmentalists found themselves opposing a native people in a land use issue, as with the Pueblo of Sandia. David Henderson, the National Audubon Society's New Mexico representative, was quoted as saying: "I've supported the commission because I'm a professional advocate for wildlife, but we've made some myopic decisions in our desire to protect wildlife. We haven't considered the human factor as much as we should. Maybe there should be more flexibility."[35]

In an area of the Carson National Forest south of Tierra Amarilla, the Vallecitos Federal Sustained Yield Unit, 73,600 acres of mostly ponderosa pine with some Douglas fir and spruce, was set aside by the federal government in the 1940s ostensibly to protect timber as a renewable resource and to help stabilize the economies of the small, forest-dependent communities adjacent to the unit through the sale of forest products. Yet conflicts continually arose between the communities, the Forest Service, environmentalists, and Duke City, the largest logging company in New Mexico, which in flagrant contradiction to the intent of the Sustained Yield Unit was doing all the cutting or contracting to local loggers. The local mill opened and closed with the fluctuations of a timber industry that was in trouble because of too much indiscriminate cutting for too many years, the obsolescence of saw mills, less reliance on lumber products for building, and the exportation of timber. But a group of local residents, some of them former sawmill workers, formed a wood products cooperative that concentrated on other skills and potential areas of employment as an alternative to wage labor at the hands of Duke City.

Madera Forest Products Association, started in 1988, included membership from the largely Hispano villages adjacent to the Sustained Yield Unit—Vallecitos, Cañon Plaza, La Madera, Petaca, Las Tablas, and Servilleta. For a time, 50 people from these poor communities were employed by the sawmill; Luis Torres, director of Madera Forest Products, hoped that perhaps twice that number would be employed in economic development projects that included the manufacturing of *vigas*, *latillas*, cabinets, and kiva ladders. Torres explained that while many people think that loggers do not want to be retrained, "it is the dependence that comes after many years in that kind of business," and seeing everyone around them losing their jobs that "they are going to try to hold onto that job."[36] The loggers were caught in the middle, he said, forced into a position that sustains the community but is not necessarily sustaining the timber base.

Luis Torres

Luis Torres was politicized as a student at the University of New Mexico in the 1960s, along with many others. But unlike most of them, after 30 years in the "real" world he remained an activist and community organizer, working with northern New Mexico communities to diversify economies traditionally based on logging. He knows a lot about logging, having grown up on a ranch in Colfax County where he worked with his father as a woodcutter and later as a Forest Service employee. He also knows that unless the loggers of northern New Mexico and the Forest Service work together to both reduce some of the proposed timber sales in the Vallecitos Sustained Yield Unit and keep the wood in the community to be later sent out as cabinets, corbels, latillas, etc., the Forest Service is going to continue to be plagued with appeals and the workers are going to be without a livelihood. "The Sustained Yield Unit has never evolved or really done anything for the progress of laborers," he says. "By now—40 years after the unit was set aside—the loggers and mill workers should be in the management organization, not getting laid off when the mill shuts down. At some point, Duke City has to get out of there."[37]

During its first year, Madera Forest Products created 300 employment days; it wanted to increase that number to expand sales to $250,000, but as Torres pointed out, the biggest obstacle to that goal was the lack of equipment that is basic to a forest products organization. In the past, members of the association were employed by Duke City or others and never acquired their own commercial saw equipment or trucks. The association tried to raise the money to invest in heavy equipment to handle *vigas* and heavy-duty chainsaws and splitters to replace hand tools, anticipating that half their income derived from fuelwood sales and the other half from forest products such as *latillas* and *vigas*.

While Madera Forest Products had no formal ties with the government, it developed contracts with the Forest Service for thinning and erosion control. Through a special-use permit, the Forest Service loaned the co-op three acres of land and a cabin to be renovated for use as its administrative office. Known as Borracho Cabin (named for a drunk steer that died in a nearby cabin—why the steer was drunk or in a cabin is a mystery), this historic cabin was originally one of the first ranger stations in the Vallecitos area but was eventually abandoned and fell into a state of disrepair. Assisted by a historic preservation officer of the National Park Service and several Forest Service archaeologists, the co-op restored the cabin to its original condition.

Unfortunately, Madera Forest Products fell victim to Forest Service policies that ran counter to rural development goals: no line item budget for rural development; a bias towards large timber sales and against small sales and nonprofit contractors; lack of technical assistance and insensitivity to local conditions at the ranger district level; and national timber targets that overruled local decisions. By the mid 1990s the organization had pretty much ceased to exist as the battles over the Vallecitos Unit resources escalated into full-blown war, which will be explored in ensuing chapters. But while Luis Torres moved on to other projects, Manuel Gurulé, a longtime logger from Vallecitos, took the reins, and with the help of Maria Varela and the infusion of grant money from foundations supporting rural initiatives, Madera Forest Community Association, as it was called, was given new life in 1998. They purchased a church site in Vallecitos to house their newly formed business assistance center and hired staff to oversee marketing programs for sustainably harvested fuel wood, traditional and contemporary wood furniture, and traditional arts and crafts.

Escalating battles over control of forest resources also worked to subvert the good work of Ganados del Valle and Madera Forest Products. Much of the controversy took place on the east side of the Carson National Forest, particularly on the Camino Real Ranger District (headquartered in Peñasco) where La Comunidad, a locally-based environmental group, challenged timber sales. One of the main movers in La Comunidad was Joanie Berde.

Joanie Berde

Berde has lived in northern New Mexico for over 30 years, mostly in Llano, a beautiful village above the Rio Chiquito outside Peñasco. But it wasn't until she stumbled across the ravages of the Alamitos timber sale on the north side of the Pecos Wilderness that she became a forest activist. It was too late to do anything about the Alamitos sale—clearcuts, silt-filled streams, eroded hillsides—but she vowed she would do everything in her power to prevent what occurred there from happening in the other Carson National Forest timber sales. As a member of La Comunidad and Carson Forest Watch, she's been involved in appeals of the Alamo-Dinner and Felipito sales, and is a constant thorn in the side of the Forest Service, lobbying for a forest plan update, moratorium on old growth cuts, and the listing of the Mexican spotted owl as a threatened and endangered species.

She's also stepped on many of her neighbors' toes along the way. While she says she's supportive of small, local wood products efforts as opposed to the large acreage cuts of Duke City, when even those cuts occur in mature ponderosa or roadless areas or in owl and goshawk habitat, she wants them shut down.

"I couldn't do the work I do if I hadn't lived here as long as I have. While there aren't that many local people who are loggers and thinners, like over in Vallecitos or in so many of the small Oregon towns, some local folks see it like the Forest Service wants them to see it—the owl or goshawk versus jobs. But really it's about new kinds of jobs—not just local logging contracts, but businesses that keep the logs in the community as vigas, cabinets, furniture, and in turn support local farming and ranching efforts. People who live here do so because it's so beautiful, and I hope they ultimately realize that something has to be done to protect that beauty and our way of life."[88]

Carson National Forest

"Just as, once the ground has been named, a woman can no longer remain completely merged with it, so too, a man, once the ground has been named as a living reality, can no longer remain so comfortably split from it."

—Starhawk

La Comunidad and Carson Forest Watch fought some big battles in the heavily logged Carson National Forest, the northern most of New Mexico's forests. La Comunidad began as an effort by some people in Peñasco to establish an ambulance service in the area, and from there went on to address a variety of issues including recycling, the Waste Isolation Pilot Plant (the repository for nuclear waste in the southern part of the state near Carlsbad), and eventually Forest Service issues revolving around logging—increased numbers of board feet, cuts on steep slopes, and clearcuts. The same Alamitos sale that attracted Joanie Berdie's attention got La Comunidad's as well in 1989, as local people witnessed a demonstration of Forest Service practices at their very worst. Located off Forest Road 161, which accesses the Pecos Wilderness from the north, the destruction the loggers wrought was clearly visible from the forest road—clearcut hillsides, streams diverted from their natural course to run alongside logging roads, slash and forest litter everywhere. Cuts like the Alamitos sale are usually hidden from public view behind buffer zones next to highways and access roads, but the Alamitos sale lay blatantly visible alongside a heavily used road and cross-country ski area (I imagine who ever designed the sale has heard plenty about his failure to hide it).

La Comunidad instigated a barrage of letters complaining of the sale to the New Mexico Congressional delegation, and Berde even traveled to Washington, DC with photographs she had taken of the sale to show Senators Pete Domenici and Jeff Bingaman, and Representative Bill Richardson. The delegation could hardly ignore the devastation depicted in the photos, and letters were written to the Forest Service complaining of what the congressmen saw, asking for an accountability of the kinds of practices employed in the sale.

Unfortunately, accountability is often only a Forest Service practice on paper. While environmental analyses and impact statements are used to justify management practices by supposedly following NEPA guidelines, when timber administrators, silviculturists, biologists, and district rangers are asked to account for what actually happened on the ground, they aren't there. In typical Forest Service rotation, they've already moved on to a new district, a new forest, and a promotion, up the ladder of constant vertical movement that defines career success. While the Camino Real Ranger District, which administered the Alamitos sale, eventually admitted the sale was mismanaged, no personal culpability was evidenced that might prevent the same kinds of destructive practices employed by the people in the actual on-the-ground management.

The next sale that came up on the district showed little improvement, at least in the planning stages. Originally slated to harvest 8 million board feet, the Alamo-Dinner sale was withdrawn and rewritten to address the needs of the local community through small sale programs. La Comunidad, the Las Trampas Land Grant, and the local Hispano community were all involved in the public input process to determine the size and cutting areas of the new sale. Through meetings with these groups, and individual home visits in the Peñasco area, the Forest Service designed the management alternatives for the sale to meet small commercial and personal firewood use. Unfortunately, when the Environmental Impact Statement (EIS) was released in 1991, because of dictated national harvest quotas, levels were set much higher than the Forest Service had indicated (5.5 million board feet), threatening old growth ponderosa pine and goshawk habitat, according to Berde. La Comunidad was especially disheartened by the EIS, as the group felt that by slating the sale for personal firewood use but keeping the harvest levels high the Forest Service was trying to buy off the local community while doing nothing to protect the old growth and wildlife habitat. While the Las Trampas Land Grant initially opposed the EIS, as debate over the sale dragged on the land grant's interest became more directed toward the rewriting of the Region Three Policy Report (*The People of Northern New*

Mexico and National Forests), a Forest Service policy written in the late 1960s at the behest of then Regional Forester William Hurst, addressing some of the concerns of rural northern New Mexico in response to the activities of the Alianza Federal de las Mercedes and La Raza Unida, which culminated in the raid on the courthouse in Tierra Amarilla (see La Raza chapter).

Because of the divisiveness in the community created by the EIS, La Comunidad decided not to appeal the sale. Under the auspices of Carson Forest Watch, however, Joanie Berde appealed it (as an intervener), already under appeal by the Santa Fe environmental group Forest Guardians. The appeal was denied, and according to Berde, the sale was mismanaged from the start: more firewood permits were issued than demand warranted; the minimum bid price on the small sales were too high for most of the local bidders to afford; bids were opened in the fall so that bidders' money would be tied up until spring; and roads into some of the sale areas were not maintained so that local people could access them. The Forest Service eventually offered the remaining uncut marked timber to Duke City Lumber.

Camino Real Ranger District personnel evidently learned from all this controversy, however. Their efforts to be more responsive to communities in the Alamo-Dinner sale grew into a strategy they labeled "Collaborative Stewardship." According to M. A. "Crockett" Dumas, district ranger, this plan was the result of district managers realizing in the early 1990s that implementation of forest plans was not working as projected. The plans too often failed to address the needs and desires of the people who use the forest and often ended up being litigated in court. Dumas and his staff, most of whom live in the local communities, determined to spend more time in the field and less in court and initiated the door to door survey of residents in the communities immediately adjacent to the forest to find out what their priorities were concerning the forest and how they hoped to realize them. As a result, they divided the district into nine zones based upon topography, vegetation, and social influence. They then mapped the areas' existing conditions with regard to watersheds, fisheries, roads, vegetation types, vegetation conditions (young, middle-aged, and old), wildlife, fire concerns, customs and traditions, and recreation. This gave them a picture of what the areas looked like at the time, and how the people used them. Next, factoring in the information they had gathered from the survey, they projected the desired condition, i.e., what the people want the area to look like and how they would like to use it. And last, they projected what processes they could use to attain the desired condition. Dumas and his staff emphasized that none of this was "cast in cement." They viewed these management strategies as "working documents" which are always open to review and revision.

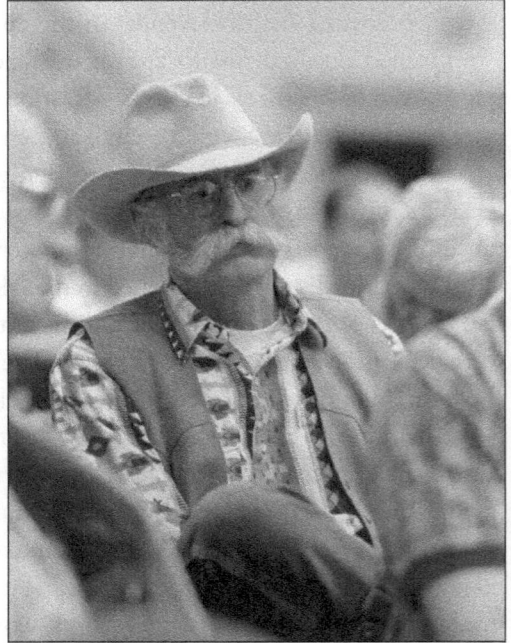

Crockett Dumas.
Photograph by Eric Shultz.

Crockett Dumas

In many ways Crockett Dumas seemed one of the least likely Forest Service employees to ever initiate substantive change. Dressed in ten-gallon hat, knee-high riding boots, and handlebar mustache, his image spoke good-old boy. Conversationally, he liked to talk a lot about his ranch in Utah where he planned to retire, and his twenty-horse endurance race stock that he and his wife competed with all over the west. But nearing the end of his career, and faced with a never-ending string of lawsuits by environmentalists and constant battles with community members, Crockett devised a new strategy. Over the course of the next three years, he worked the district, connecting with land grant people, local environmentalists, with Picuris Pueblo, from door to door. His listening skills improved while his rambling conversations became more focused. Recognition from the forest supervisor's office, then from the Washington bureaucracy, that Crockett might actually be on to something, extended the collaborative program to other local districts as well as all over the west. Crockett, with community members in tow, traveled to Montana and Oregon to present his case. To end his career with a resounding pat on the back rather than a count down of his days left was his scenario of choice.

Carson Forest Watch also appealed the Felipito timber sale in the Vallecitos Federal Sustained Yield Unit, a 4.1 million board feet sale. The two main issues in the Felipito sale, according to Berde, were spotted owl habitat in potential old growth, and the depletion of renewable timber resources that would eventually affect the local community. La Compania Ocho, a Vallecitos-based logging group (the subject of the next chapter), had already been asking the Forest Service to rewrite the management plan for the Vallecitos Unit, citing that the plan was outdated and timber inventories were inaccurate. Carson Forest Watch's concern was that potential old growth acres were included in the sale, and that, ideally, these acres should be designated old growth even if they didn't meet all the requirements for that classification (some of the land was logged at the turn of the century).

Despite the fact that the U.S. Fish and Wildlife Service, the New Mexico Department of Game and Fish, and many of the local people were opposed to the size of the sale, Carson Forest Watch's appeal (Sam Hitt of Forest Guardians was an intervener in the appeal) was denied, and a discretionary review by the Washington office was rejected. Berde was especially concerned that while Carson Forest Watch considered a lawsuit (it was eventually filed), roads into the Felipito sale would cause irrevocable damage and open up a forest ecosystem to poaching, firewood, and ORV use: "Once you put a road in—I've seen it happen a million times—you lose it. The intactness of that sort of forest just goes.... They never close the roads like they say."[39]

When I interviewed Berde about her group's appeal of the Felipito, she mentioned that several other environmental groups had refused to join the appeal because they saw it as an "owl versus jobs" issue and didn't want to alienate local communities. When I told her I would also be interviewing Luis Torres of Madera Forest Products, she said, "Yeah, he probably hates me."[40] Subsequently, in my interview with Torres, he expressed his mistrust of Berde as an "outsider," not sensitive enough to his people's economic dilemma. He also pointed out that it was not an issue of "owls versus jobs" but an issue of sovereignty.[41]

Unfortunately, as I learned in the battles in Placitas, there is a very real distrust between the indigenous population and the Anglo environmentalists who so often have an urban orientation and little sensitivity to rural issues. As I also learned, the Forest Service has used this very distrust to divide and conquer its opposition. For example, in a timber sale summary for 1992, Carson National Forest claimed that 10 million board feet of timber was either delayed or not offered because of the Mexican spotted owl. It's the same thing the Forest Service did in the Northwest with the northern spotted owl to fuel the fires between the loggers and environmentalists, ignoring the real issues of depletion of timber resources, outdated lumber mills, and exports. Here in New Mexico, timber sales were being delayed or not offered for many of the same reasons; Berde claimed the report by the Carson was a political move to pressure the U.S. Fish and Wildlife Service not to list the Mexican spotted owl as an endangered species (the owl was listed as a threatened species in March of 1993). Berde's position was that it's sometimes necessary to refuse to compromise with the Forest Service when old growth and potential old growth forests are at stake. But by taking this hard line she ended up fighting the very people who have been the vanguard in the battle to kick the corporate timber interests out of northern New Mexico. Environmentalists and the people who live and work in forest communities should be *vecinos*, not adversaries, working to sustain our timber resources and way of life.

La Raza

"You can shut down the woods but we still aren't leaving. We were here *before* there was timbering."
—Antonio "Ike" DeVargas

Antonio "Ike" DeVargas is a *vecino* who rarely sees eye to eye with Berde and isn't afraid of assessing blame wherever he sees it. Knowing of his long-time involvement in land grant and forest issues, I looked forward to interviewing him.

Mark, Max, and I drove over to the small village of Servilleta Plaza to meet with DeVargas in the same house where he was filmed twenty years ago by underground filmmaker Danny Lyon. In "Little Boy," Lyon's movie, DeVargas is featured prominently as a founding member of New Mexico's La Raza Unida party, dealing with the same issues he's always been involved in— land use, forest use, and the economic stability of Hispano rural communities. Intense and articulate, DeVargas was an effective spokesman for his cause. Twenty years later at our meeting, his curly black hair completely white, his green eyes surrounded by wrinkles, he was just as emphatic.

Servilleta Plaza is a village of five families in a valley of piñon/juniper vegetation on the west side of Carson National Forest. While Mark and I had once looked at a house for sale in the nearby village of La Madera, we'd never before been to Servilleta Plaza, or even heard of it, for that matter. This west side of the Carson seems even more remote than the east side (of the Rio Grande, which bisects the forest) where I live, further away from Taos and Santa Fe, accessed by only recently paved roads. And as part of the Vallecitos Federal Sustained Yield Unit, it has been heavily logged for years.

We found DeVargas's house with no trouble (actually a neighbor's house, as he said his trailer had yet to be hooked up with utilities), and were met by a huge black mastiff. DeVargas led us into the living room of an old adobe, adorned with family pictures and religious icons. I recognized a picture of one of the Head Start teachers whom I had met recently at a workshop for parents.

DeVargas was extremely polite and accommodating, knowing nothing of me other than I was in the process of writing a book about forest and rural issues. While Max entertained himself with toys he found in a box behind the couch, DeVargas told us a little of his background. He grew up in both La Madera and Española with extended family members, and as a young man worked in the local forests hauling firewood, thinning, and logging. Like many young men in northern New Mexico, he dropped out of high school, joined the army (or in Ike's case, the marines), and was sent to Vietnam, and like many of these same men, his political education crystallized there: "Most of us went there believing what the government told us, that what they were doing over there was good and necessary, and most of us came back knowing that if they were lying to us over there they were lying to us here, too."

Antonio "Ike" DeVargas.
Photograph by Eric Shultz.

DeVargas spent five years after his Vietnam tour traveling around the country, trying to adjust to being back in the States, trying to find his bearings. He said, "I finally realized I had to be in the woods to survive," and he came back to New Mexico, specifically Servilleta Plaza, where he quickly became involved with La Raza Unida, a Chicano activist group organized at the national level but with a large membership in New Mexico. While DeVargas was still in Vietnam, most of the local attention had been focused on another northern New Mexico activist, Reies López Tijerina (Tijerina is originally from Texas), whose Alianza Federal de las Mercedes, leading the fight for the readjudication of land grant claims, had been involved in the infamous raid on the Rio Arriba County courthouse in Tierra Amarilla. DeVargas explained that the two groups, the Alianza and La Raza, were involved in different issues, and his personal experience with Tijerina had not been good. After Tijerina came out of prison (he served several years in prison for assault on a federal officer and destruction of federal property), La Raza invited him to give a speech at one of their meetings and Tijerina had not been supportive.

From 1980 to 1982 DeVargas worked at La Clinica in Tierra Amarilla (where Maria Varela also worked), and as a member of the Vallecitos Association, a community advocacy group, became involved in forest issues when the Carson Forest Plan, promulgated in 1985, set the timber harvest levels for the Vallecitos Federal Sustained Yield Unit. "We told them right off that their ASQ [Allowable Sale Quantity] was not sustainable," DeVargas said. The association felt that a volume of 3.5 million board feet (mmbf) was a reasonable cut in the Vallecitos Unit, while the Forest Plan called for 8 mmbf. An agreement was finally made that assured 5.5 mmbf to Duke City Lumber, the largest commercial timber company in New Mexico (and contractor of almost all small logging companies), with 1.1 mmbf of sawtimber set aside for small loggers and 1.1 mmbf set aside for wood products such as *vigas* and *latillas*.

After the Plan was implemented, DeVargas became involved with Luis Torres and Madera Forest Products Association, which, as I documented in the "Norteño Life" chapter, was mainly involved in trying to establish wood products businesses to replace the sawtimber business. As a community member dependent upon logging for his own livelihood, DeVargas soon realized that things were moving too slowly with the non-profit Madera Forest Products to provide immediate relief and long-term stability for local loggers. Because the local mill in Vallecitos continued to open and close due to fluctuations of the timber market, the controversies surrounding the spotted owl, and the effects of years of over cutting in the Vallecitos Unit, DeVargas decided to organize his own for-profit company that could compete with Duke City for sawtimber and whose members could have more control over their own destinies. He felt that with a sustainable cut set by the Forest Service and the joint effort of Madera's wood products businesses and his company's sawtimber business, the local industry could be kept viable and the community employed. In 1990 eight members invested $1,000 each, and with a loan guaranteed by all their combined land and possessions, bought enough equipment—loaders, skidders, trucks, bulldozers—to start La Companía Ocho. "I'd like to see Duke City kicked out of the Carson and Santa Fe National Forests by 1995," DeVargas said. "This is another pendulum swing of the clock—from the small communities to the big corporations back to the small community mills. There will always be a need for timber management, without corporate timbering, and that's where we fit in." DeVargas and members of the Vallecitos Association began to rewrite the Vallecitos Sustained Yield Unit Plan to reflect current needs and sustainable management.

Right before I saw DeVargas, he had been interviewed for an article in the *Albuquerque Journal* called "Forest Families: Rooted and Uprooted." "I told the reporter, it's not really worth it to invest any more money in La Companía because we're not getting the raw material. The Forest Service is not selling us the timber we need. The 1.1 million board feet they promised they've never delivered." In the next paragraph, the spokesman for Carson National Forest said that the agency tried to offer the wood, only to be stymied by appeals from environmentalists. Knowing that there was no love lost between DeVargas and some environmentalists (he had recently organized a demonstration at environmental lawyer Grove Burnett's private ranch outside the town of Vallecitos, protesting a gathering there of national environmental activists), I asked DeVargas who the real bad guy was, the Forest Service or the environmentalist.

He immediately said it was the Forest Service, in collusion with the big logging companies, that was responsible for the over cutting and the lack of support of local community needs (La Companía was, at the time of the interview, involved in a dispute with Duke City over the terms of its contract to log a sale in the Vallecitos Unit because the company was offering

La Compañía less money than it did the year before to log the same area). But he minced no words about environmentalists, either. "Environmental groups' arguments are not based on reality. They're using the spotted owl, which isn't here, to halt timber or traditional uses of the land. They want to turn New Mexico into a museum for hikers and recreationists who don't have to live here and make a living, and the Forest Service plays up to that. I told Grove, you guys don't live here, you don't know what's going on. Don't be such elitists—put some of our people on your fancy boards." He did say that he and the people he worked with had been getting some help from Forest Trust, a nonprofit consultant group in Santa Fe, and the Green Party in Taos, whose platform encompassed both environmental and social issues instead of the single issue focus of so many environmental groups.

Earlier, when DeVargas talked about his work with Louis Torres, I had asked him if there had been any bad blood between them when DeVargas had branched out to organize La Compañía Ocho. He told me they remained friends, but that they had different styles of doing things: "We're sort of a Mutt and Jeff team, the conservative and the radical, negotiator and the stirrer-up." DeVargas sees himself as the thorn in the side of everyone who has participated, knowingly or unknowingly, in the displacement of the land-based people of northern New Mexico. It started a long time ago with the taking of community land grants and it continues today with the battles over the use of the forests and the economic viability of local communities.

Before we left, DeVargas told us of his work with Ganados del Valle, at that time pursuing a lawsuit against the Sierra Club for the alleged mismanagement of a grant dispersed through the Sierra Club but intended for Ganados (a settlement between the two parties was reached in 1995). Ganados was considering joining the Vallecitos Association in a lawsuit against the Forest Service for its noncompliance in meeting the fiduciary agreement in the Sustained Yield Unit and for the Forest Service's long-term lack of commitment to sustainable communities. DeVargas was getting ready to make the long drive to Tierra Amarilla to meet with Ganados that afternoon. "You know," he said, "logging isn't an easy job. I'd gladly give it up if there was anything else I could do. But that's what this is all about, getting the powers that be to recognize that we're always going to be here and we have to make a living. You can shut down the woods, but we still aren't leaving. We were here *before* there was timbering."

(La Companía Ocho and the Vallecitos Association filed a lawsuit against the Forest Service in U.S. District Court on March 25, 1994 for failure to meet its fiduciary responsibility in the Vallecitos Federal Sustained Yield Unit. A settlement was reached in February of 1995 that guaranteed La Companía 75 percent of the sawtimber from the 2.1 million board feet in the La Manga timber sale and 80 percent of the sawtimber from the Agua/Caballos sale, without competitive bidding.)

The Real Work

"The problem is, where do you put your feet down, where do you raise your children, what do you do with your hands?"

—Gary Snyder

Note: This is a piece I wrote soon after we moved to El Valle. I'm including it as written, in the present tense, for its first impression of and immediacy to our new home.

Trailheads leading to the Pecos Wilderness are fifteen minutes from our house. Alamo-Dinner Canyon, site of the Carson National Forest timber sale, lies just over the northeast ridge from our valley. On the other side of the valley, second-and-third-growth ponderosa pine has finally filled in the bare spots of decades old timbering.

We are here, in the heart of the Pecos country, doing what everyone else does to survive. As I write this, on a warm March day of false spring (we all know winter is only on a brief vacation), I can see the still significant remains of our woodpile in front of the house. After twenty years of heating with wood, Mark still enjoys cutting wood and collected our pile with a vengeance last summer and fall. In Placitas we were spoiled, having had access to cleared grazing land that provided piñon and juniper fuelwood. Up here, we have mainly the softer woods—aspen, ponderosa, spruce—but plenty of it, the detritus of winter's wind and snowfall that we gather forest wide, or marked timber from designated Forest Service cutting areas. Permits are issued for a minimal amount, and the morning I showed up at the ranger station at 6:30 a.m. for one of 75 special permits, there was already a line.

We're still doing the same kinds of things we've been doing for twenty years to make a living, a kind of mixed bag of various cottage industries coupled with a teaching gig here and there. Mark started teaching the local GED program, and I'm still writing guidebooks for backpackers and cross-country skiers. For several years now we've been selling garlic, sweet peas, and shallots to markets in Taos and Santa Fe.

Garlic grows well at 8,000 feet.

Our first winter in northern New Mexico was one of the wettest on record. And except for the week of rain over Christmas, and another week of rain during the February thaw, the skiing was great. We finally bought skins and cable bindings for our cross-country skis to climb into the Pecos backcountry, and we downhill skied for the first time in twenty years at the neighborhood ski area, Sipapu. Kids from the local schools ski for a reduced rate every Monday for eight weeks, and their parents can come, too, for the same price. I must admit, the downhill thrill proved to be as insidious as ever, and we enjoy our Monday afternoons flying down the hills. Unfortunately, Sipapu, essentially a local draw with some Texans thrown in for profit, applied for a permit to expand the area, increasing its vertical drop by one hundred percent. While the proposed expansion hasn't elicited the kind of organized, public opposition that the proposed Santa Fe Ski Area expansion has, many local people, some of whom work at Sipapu, worry that the expansion will alter Sipapu's commitment to the community, contaminate the Rio Pueblo, which runs through its lands, and increase the chances of local development. When a committee was formed, organized by Carl "Cat" Tsosie of Picuris Pueblo, to see what could be done to halt the expansion, Mark and I were asked to join.

Our bimonthly meetings (we have the unwieldy name Rio Pueblo/ Rio Embudo Watershed Protection Coalition) attracted a diverse group of people, concerned about their river, the Rio Pueblo, its tributaries, and their communities: *acequia* associations from Vadito, Embudo, and Dixon; the Eight Northern Pueblo Council; land grant activists; Carson Forest Watch; the Sierra Club; former Sipapu employees; and representatives of the Mora Water and Land Protective Association, who were concerned expansion would affect their northeast side of the Sangre de Cristo Mountains as well. We filed an administrative appeal of the Sipapu Final Environmental Impact Statement (the ski area has a permit to operate on Carson National Forest land), released in March of 1995, which approved everything the ski area wanted: an increase from 185 acres to 977 acres; a doubled skier capacity; two more ski lifts; a second restaurant/ski patrol building at the top of the mountain; more parking lots; new motel units. In June, the Forest Service withdrew its Record of Decision approving the Sipapu expansion, admitting that it had failed to properly analyze cultural properties in the area, which encompasses Picuris Pueblo ancestral lands, as stipulated by the National Historic Preservation Act (reminiscent of the decision to halt road development in Las Huertas Canyon in the Sandias). We also filed a protest of a water rights transfer requested by the ski area to increase its snowmaking activities (the water rights the area claimed it possessed were under question as well). But the meetings provided more than working sessions on how to file a Forest Service appeal—they reinforced what we'd been learning in our own little village, that there is indeed a community here in northern New Mexico that is very protective of its way of life.

Mark and I wrote our first ever grant proposal, asking for money to underwrite a community newspaper that would cover the issues the coalition was dealing with as well as all the other issues concerning resource and cultural sustainability. We had thought about publishing a newspaper for years, back in Placitas, to join our writing skills with our activism, but the project was somehow lost in the struggle. Much to our surprise, our first proposal was funded, and with several thousand dollars we began publication of *La Jicarita News* (Jicarita is the name of the mountain peak at the head of our watershed), which is mailed to an extensive list of subscribers and distributed throughout northern New Mexico.

Our first Christmas was cold and white, particularly the night of the village Las Posadas. The parish priest visits a different village every night, ending the celebration on Christmas Eve at the Santuario in Chimayó. Placitas always celebrated Las Posadas as well, and over the years we were there for many, one time loaning Raisin, our goat, to guard the manger. A procession through the village stopped at designated houses singing, "Will you let us in," only to be turned away, until refuge was found in the school gym, where a manger scene and party culminated the celebration. Las Posadas in Placitas grew each year as the area grew, and more people were curious to see this tradition that not so long ago was almost forgotten in the rush to make all things modern. Here, in our northern New Mexico Posada, many of the same traditions are observed: two school kids dressed as Mary and Joseph shyly lead the procession past brightly lit *luminarias*; the guitars and Christmas carols swell the walls of the closely packed church; the food and good spirits are shared by all who care to come.

On Christmas Day we ate dinner with our neighbors, an extended family of grandfather, Tomás Montoya, uncle, aunt, son, daughter, grandchildren, and assorted mates, lovers, friends. I contributed a cherry pie (with sour cherries from the trees outside our front door) to the feast of roast beef, *tamales*, *chicos*, mashed potatoes with red *chile*, salad, vegetables, and homemade bread. We, residents of six months' duration, were invited into the home of these people who had already demonstrated their neighborliness by watching our kids, helping install our wood stove, and plowing our driveway after every snow storm.

We finally invested in a heavy-duty rototiller so we could double the garden size to accommodate more corn, peas (they grow all summer long, never bolting), potatoes, and even a few yuppie greens. The garlic was in by October—six different types in an attempt to find the perfect match for soil and climate. They're all big, juicy, and hot, and we've been able to sell everything we've grown in the Santa Fe markets. We'll never grow on a scale that makes us more than gentlemen farmers, but our hands will be dirty and our food will be fresh.

Most of our neighbors grow hay for their cattle, which they've done for many years and are beginning to grow vegetable gardens again, as their parents and grandparents did before super highways brought super markets in Taos and Santa Fe within commutable distance. Corn, squash, and *chile* are staples, as they've always been, but Gerson, one of our neighbors, says he's planting pinto beans, peas, and carrots this year. Retired from the Las Vegas mines, maybe he'll finally have the time to grow a real garden, fence in his pasture, and hike to the lakes.

We all go to the Pecos lakes that head our valley, either on foot or horseback, with fishing pole or backpack, to hunt with rifle or camera, to lift our spirits, get rid of a little angst. Tomás and his son Fred go to look for cows; Rico goes to ski the glacial chutes that hold snow until June; I go just to go—to walk, stretch, reach, and breathe. We are lucky the lakes are there, the source of our *acequias*, so close, so beautiful, so important. They're why people like Marion Davidson, Tom Nordstrom, Luis Torres, and Antonio Manzanares fight the good fight to ensure the mountains' survival and *our* survival, so inextricably linked.

I was inextricably linked to Placitas for almost twenty years. It hasn't been easy for me, breaking that bond—I miss my house, I miss the Sandia Mountains, and I miss my sense of place there, even though the last few years changed that place, and my relationship to it, irrevocably. I suffer periodic pangs of guilt that perhaps I copped out, that I left my friends in the lurch. Lizzie Archibecque will never leave, tied to the land with historical strings that only tighten with age.

I console myself that I've changed only my neighborhood, not my connection to life in the high desert, a bioregion that includes Placitas *and* northern New Mexico. Instead of the Sandias, I have the Sangre de Cristos. Instead of one adobe house I have a different adobe house, worked on with the same loving care and attention. And instead of subdividers and yuppies for neighbors, I have neighbors who are trying to make better choices. In his book *Real Work* Gary Snyder talks about "Right Livelihood. Work that doesn't cheat people.... But we feel bad because we find ourselves doing things which are implicitly valid but are hooked up somehow to the economic growth system which is out of control." There is a difference between pursuing a sustainable life and that which consciously "cheats" both people and the land. You can be a logger without working for Duke City Lumber. You can be a builder without working for AMREP or the Placitas Land Company. You can use a computer without buying into a technology that refuses to acknowledge that the "biological world is infinitely more complex" (Gary Snyder). And you can hopefully move to a new community without being an invader. You can try to be a sustainer—part of the solution, not the problem.

Las Sierras

"What's being done here is like what multinational companies do in third world countries: use up the resources and then move on. This is the heart of the issue that makes people here so angry. We feel exploited for economic gain. It's the mentality of take, take, take without giving back to the people and the land."
—Greg Cajete, Santa Clara Pueblo

Once I became a resident of Taos County, and traveled to Taos at least once a week to do my shopping and take care of business, I become familiar with the town's skirmishes and battles as a growing tourist haven even if I didn't take an active role in the war. And there have been many battles over the years—the proposed expansion of the Taos Municipal Airport, Taos Golf Properties, and, of course, Las Sierras. Just those two words evoke pictures of pickets in front of Coldwell Banker Lota Realty and crude signs on fences in Las Colonias exhorting the developers to get lost.

It all began in January of 1991 when David Buck of Coldwell Banker Lota Realty announced that a consortium of California, Texas, and Chicago area investors was going to develop the property it owned and had options to buy, on Taos Mesa and in the Las Colonias valley, at the north end of Taos. This wasn't going to be just any development—according to Buck, this was going to be a project that would make Taos "the premier community in New Mexico." The plans included a 320 acre, $40 million golf course and country club surrounded by 150 homes in Las Colonias valley, and a 216-room Las Sierras Inn, Spa, and Conference Center, with a nearby attached commercial center on Taos Mesa, the dry, sagebrush covered *llano* above the valley. The developers had already applied to the El Prado Water and Sanitation District, which governs water in this northern part of Taos, to transfer 140 acres of surface water rights from the Las Colonias property where the golf course was planned to its well in Upper Las Colonias, from which water for the development would be pumped. Estimated consumption for the development would be 400,000 gallons of water a day at its peak. Buck claimed that the project was fully funded, and if water transfer rights were approved, the project would begin that summer. He also anticipated some Taoseños would oppose his project, and was quoted in the *Taos News* as saying, "If it becomes too ugly, they'll [the investors] probably reconsider doing it at all. If it takes years, we might not do it. This is, we think, the cleanest, most effective way to do something for Taos County."

I'm sure David Buck often ate his words over the course of the next two years, as Las Sierras became the most controversial issue of 1992, and by 1993 was still a figment of his fertile imagination. The proposed development generated instant opposition, of course: Las Colonias-West Mesa Preservation Association, which had been founded several years before to deal with previous threats to the communities, regrouped and elected JoAnn Medina, a native of Las Colonias, and Kat Duff, a nine year resident, as co-chairpersons. The group immediately filed a protest of the water transfer request with the Office of the State Engineer (OSE); in February, a demonstration in front of Coldwell Banker Lota Realty confronted David Buck with signs like "Coldwell Banker takes $ Gives Back Sewage." Residents also appeared at a county commission meeting to ask the commission to impose a moratorium on any further development in Taos County until a land-use plan was in place. Several people pointed out what they saw as a conflict of interest between members of El Prado Water and Sanitation District, John Painter and Telesfor Gonzales, and the developers: Painter worked for David Buck and Telesfor owned land where the sewage treatment plant was proposed. The commission, reluctant to get involved in zoning issues, ignored the citizens' request. When Las Sierras submitted subdivision plans to the county in the summer of 1992, the commission, adhering to subdivision regulations, couldn't act upon the development until various agencies such as the OSE, the Environmental Improvement Division, and the Soil Conservation Service looked at the various aspects of the proposal and made recommendations.

Letters and guest editorials appeared in the *Taos News* delineating opposition to the development (only about half of those submitted were ever run). The water transfer plan was of major concern, of course: A water transfer would dehydrate an existing agricultural base, could leave residents with water shortages during late summer, early fall, and could require residents to upgrade their wells. Opponents to Las Sierras hotly contested Buck's claim that the proposed development would use only 400,000 gallons of water, citing that a golf course alone might use 1,000,000 gallons per day. Outrage was expressed concerning the destruction of agricultural and wild lands for houses and a playground for the rich, menial jobs at minimum wage as what this "trickle down" development had to offer, and increased land values with local property owners facing double and triple land taxes. The developers immediately labeled the opposition "anti-growth," as they so often do in attempts to deflect real concerns about what is happening in New Mexico.

In March of 1992 David Buck sent a letter to various Taos businesses challenging opponents' claims. To dispel any notions that this was a development for the rich, he pointed out that the golf course would be open to the public, as would the performing arts center. The sewage treatment plant would be state of the art and would take care of the pollution problems created by septic tanks (the local people weren't aware they had any). Water would be taken from Upper Las Colonias wells in El Prado Water and Sanitation District to lessen the impact on one aquifer. Public transportation in the Taos area would be enhanced by shuttles from the development to downtown and the Taos Ski Valley. Ninety percent of the estimated 220 jobs created would be reserved for the locals and only 35 of those jobs would pay less than $15,000 (he never said how he arrived at these figures). Increased property taxes would be offset by huge new county tax revenues, and besides, it would be easier for local property owners to rent, sell, or ask for loans.

Buck's statement clearly revealed the crux of the problem. People whose families have lived in Taos or Placitas for generations, or people who have chosen to live there because of what these communities have traditionally offered—a rural, community oriented lifestyle—are not interested in "selling out." I never figured the "resale" value of my land in Placitas when I bought it, or that my purchase was any kind of "investment" other than an investment of my life to a particular community. All of David Buck's assurances that Las Sierras would be an "economic boon to Taos County," and an "absolutely top-quality" development are meaningless to people who recognize it for what it is—another rip-off of a lifestyle, the destruction of rural lands, and a fly-in-fly-out community in return.

JoAnn Medina

"The only reason I would ever leave," JoAnn Medina says, "is if I could no longer afford to stay." Born and raised in Las Colonias, where her 93-year old father still lived at the time of this interview, JoAnn worries about her children. "They're smart kids and will probably leave to go to school. But they want to come back to Taos, to live in a small town, where they were born. What will be here for them?"

JoAnn, a paralegal for more than eighteen years at the Northern New Mexico Legal Services office in Taos, has always been committed to her community—Taos and Las Colonias. At the law office she helps people apply for social security and disability benefits and works with those who want to avoid costly divorce suits and work out their own divorce agreement. She met her neighbor, Kat Duff, while working as a volunteer at the Battered Woman's Project.

"Las Colonias is a diverse community," JoAnn says, "with farmers, back-to-the-land folks of the '60s, old-time artists, and even a few very wealthy people like R.C. Gorman, in complete solidarity against Las Sierras. But the most important thing to remember about Las Colonias is that it is a community of poor people with land, and I never want to see those people lose their land."[42]

Fortunately, in May of 1992 the OSE denied Las Sierras its requested water rights transfer. State Engineer Eluid Martinez said his decision was based on the fact that a transfer might draw down area wells and might hamper the historical flow of the Rio Pueblo, the Arroyo Seco, and the Rio Lucero. In his decision he wrote, "The public welfare is not well served by approval of only a portion of the water supply required for a proposed planned development project in which the ultimate water requirements are unknown. Ultimate cumulative effects of existing water rights resulting from the use of water within the proposed Las Sierras development project are not known and therefore a determination cannot be made as to whether those effects would not constitute impairment." The Preservation Association, which had spent considerable time and money to present its case in a hearing before the OSE, was thrilled with Martinez's decision and saw it as precedent setting. For the first time in water law the notion of "public welfare" was used to evaluate the cumulative effects of a proposed development, perhaps limiting future developments like Las Sierras. The decision also supported traditional water uses for agricultural and religious purposes (Taos Pueblo was united with Las Colonias-West Mesa Preservation Association in its opposition to Las Sierras, worried that the water transfer would lower the water table under the Buffalo Pasture Wetlands, which contain sacred springs, herbs and clays) instead of automatically validating residential-urban uses, as so often happens in the west.

David Buck immediately declared that it was unreasonable for the State Engineer to require that a development's entire future needs be covered before construction. Somehow all the very precise figures Buck released about his development—numbers of houses, numbers of jobs, square footage of performing arts centers, salaries and monies generated—began to lose any semblance of credence when he objected to having to have a realistic idea of where all the water to support these figures was going to come from. Attorney Fred Waltz, who lived in Las Colonias, represented several local well owners at the water rights hearings and lauded the decision as one that would limit developers to reasonable projects in keeping with the character and constraints of the land upon which the projects were to be built.

With the denial of the water rights transfer, coupled with the mounting public opposition to Las Sierras (remember Buck saying in January of 1991 that if most of the county seems to be against the development the investors would not pursue it?), it seemed the appropriate time for Buck and his investors to bail out. Instead, he announced they would appeal the State Engineer's decision and would continue to explore other water options. In early summer El Prado Water and Sanitation District appealed Martinez's decision to District Judge Peggy Nelson's court. In October, Nelson dismissed the appeal, finding the district had failed to notify local *acequia* commissioners of the appeal. In February, Richard Simms, the lawyer for the water district, filed a motion for reconsideration of Judge Nelson's dismissal, alleging two errors in her procedure. Simms said if his motion was denied he would immediately appeal to the state Court of Appeals, stating that the appeal "no longer has anything to do with Las Sierras," but was an effort on the part of the water district to determine its authority to sell water. There was also talk that the district was trying to secure water rights from other sources, including the Jicarilla Apache Tribe (which had recently received water rights in the settlement of a federal lawsuit).

Suddenly, in April of 1993 Simms filed a "withdrawal of motion for reconsideration" of the water rights transfer, and David Buck announced, "we're dead in the water." Apparently unwilling to pursue more court battles, the retired veterinarian from Santa Fe who owned both the water rights in question (to be transferred under the auspices of El Prado Water and Sanitation District) and several hundred acres of land under purchase option to Las Sierras, terminated his agreement with the developers.

JoAnn Medina and Kat Duff were always confident that the denial of water transfer rights would have been upheld at all levels of the New Mexico court system and that the possibility of acquiring other water rights was farfetched. So they were happy that at least for the time being the fight was over, and remained cautiously optimistic that Las Sierras would probably not happen in the immediate future, on its original scale ("I don't even listen anymore to the figures they give on number of houses or size of golf course, they've changed so many times," Kat Duff said), and in their neighborhood. But it could still happen someplace else, in another traditional neighborhood like theirs, where once again people would have to fight to protect that neighborhood. They both recognized the need for some kind of land-use planning for Taos, although they were also aware of how difficult that would be to design and implement to protect the needs of traditional communities and provide an economic base. Soon after the Las Sierras debacle, the County Commission began visiting neighborhoods to solicit input regarding a Taos land- use plan, but the County would see many more growth battles before a plan was in place.

Kat Duff

"I owe it to people like JoAnn Medina and her family, the caretakers of a traditional community like Las Colonias, to fight for a home that is now mine as well," Kat Duff says. "And I don't want to have to leave."

Kat moved from Minnesota to Santa Fe to complete a master's degree in counseling before settling in Las Colonias in the early 1980s. A former director of the Battered Women's Project, she, too, is as committed to her adopted community as JoAnn is to her birthplace. "When I actually sat down and counted up the hours I spent dealing with the water rights transfer and development, I came up with about twenty hours a week," Kat says. That included writing grant proposals to fund the fight; the Preservation Association was fortunate enough to *receive $15,000 from several foundations. But these kinds of battles are mostly fought with "cheap woman labor," Kat jokes, and a lot of "grass roots fund raising activities like bingo, art auctions, movies, bumper stickers, etc. It's been a real victory for the people and made everyone extremely aware of all we have to lose."[43] Optimistic yet wary, Kat and JoAnn will maintain their vigilance, as all the rest of us "cheap women" continue to do as well.*

(In the fall of 1994, a group of Chicago investors, allegedly including one of the main people involved in Las Sierras, began negotiating to purchase 14,348 acres on Ute Mountain, north of Taos near the town of Costilla, for $2.8 million. A Texas man, Ray Smith, had bought the land several months before for $550,000. The investors wanted to build 470 residences, two golf courses, 500 condominiums, 100 40-acre ranches, and a large hotel. Needless to say, opposition, in the form of the Ute Mountain Protection Team, quickly rallied to protest the sale, citing the significance of Ute Mountain as an important wildlife habitat and rural landmark. As of this writing, the sale has yet to transpire).

The Mexican Spotted Owl

"Take back, like the night, that which is shared by all of us, that which is our larger being. There will be no 'tragedy of the commons' greater than this: if we do not recover the commons—regain personal, local, community, and people's direct involvement in sharing (in being) the web of the wild world—that world will keep slipping away."

—Gary Snyder

Communication between certain environmentalists and *norteños* completely broke down over the 1995 lawsuit filed by environmental groups in New Mexico and Arizona (Forest Service Region Three)—Carson Forest Watch, Forest Guardians, and Forest Conservation Council—against the Forest Service, claiming inadequate protection of Mexican spotted owl habitat. In fact, all out war was declared when Ike DeVargas, as a member of the newly formed *La Herencia del Norte*, hung Sam Hitt of Forest Guardians and John Talberth of Forest Conservation Council in effigy in downtown Santa Fe on November 24, 1995. DeVargas was quoted as saying: "It's a symbolic hanging. It's meant to put them on notice that people are getting angry enough that violence could happen."[44]

This lawsuit alleged that the Forest Service had failed to look at the cumulative impacts of logging on the Mexican spotted owl in planning its timber program. By submitting its timber projects to the U. S. Fish and Wildlife Service project by project rather than looking at the forest as a whole or forest plans as a whole, the suit claimed the Forest Service had violated the Endangered Species Act. In August of 1995, a federal judge in Arizona issued an injunction that halted all logging on the affected forests.

No one was happy with the injunction, including some of the litigants, and in October both the Forest Service and representatives of the plaintiff environmental groups sat down to negotiate a settlement. Each side had its own story to tell about the proceedings. Joanie Berde of Carson Forest Watch claimed that the Forest Service wanted only the big timber sales, such as the

Ojos Ryan and Felipito in the Carson, released by the judge, not the smaller thinning projects and *latilla* and *viga* sales the environmentalists wanted released. The Forest Service claimed that, first of all, it complied with U.S. Fish and Wildlife regulations in surveying for the spotted owl, and secondly, that Ojos Ryan (2.1 million board feet) did not qualify as a large sale.

As for firewood, the settlement placed several restrictions on the Carson. First, no dead and down wood on the forest could be gathered in the owl's habitat and in sensitive streamside zones. Maps were drawn up showing the areas where wood gathering was permitted and mailed to everyone who had obtained a firewood permit. According to the environmentalists, critical owl habitat amounted to less than a third of the area within a 20-mile radius of Peñasco. Second, the only cutting allowed of green trees was from marked areas. Standing dead piñon and juniper could be cut outside critical habitat, while ponderosa pine could not.

For many years (since the early 1970s with only a few restrictions in place) the Carson had an unrestricted firewood gathering policy that allowed for free wood gathering anywhere on the forest and for the cutting of standing dead trees. According to Berde, the Forest Service recognized there was a problem with this policy because of declining populations of songbirds and other wildlife that depend upon dead and down trees for habitat. But because of anticipated backlash from the community, the Forest Service used the settlement negotiations as a vehicle to change its firewood program, thereby making it seem like it was the environmentalists who were denying people wood. While Berde and other environmentalists agreed that there needed to be reasonable restrictions on the firewood policy, they felt the Forest Service should have done it the right way with public input and amending the forest plan. They also accused the Forest Service of not offering enough thinning sales in low impact areas for people to cut green wood instead of emphasizing large commercial sales.

DeVargas was upset by the settlement because he felt it not only failed to address the needs of the Vallecitos community, which is supposed to benefit from the Sustained Yield Unit, but that it was preventing him and his coworkers in La Compañía from getting back to work. La Compañía Ocho had been trying to cut the La Manga timber sale, the newest sale in the Unit, which Forest Guardians and Forest Conservation Council had already appealed, citing lack of protection of old growth. DeVargas and La Compañía had joined the appeal as an intervener, citing economic interest. In September of 1996, a federal judge ordered the Forest Service to offer the sale to La Compañía, stating that the injunction halting all logging in Region Three did not apply to La Manga. The environmentalists immediately went back to the judge asking for a stay on his decision until there was clarification as to whether the logging injunction applied to the sale.

DeVargas called the environmentalists "interlopers" and claimed they wanted the forests managed like those in Oregon or Washington when these New Mexico forests have a unique history of Native American and Hispano land grants, as well as the special Sustained Yield Unit designation. He again called for the implementation of The Region Three Policy Report, which mandated that the villages and pueblos of northern New Mexico be treated as a natural resource, with their welfare foremost in mind.

More controversy developed on our side of the Rio Grande when the Truchas Land Grant declared that because of firewood restrictions many people in the village of Truchas did not have enough wood to get through the winter. Because the Truchas Land Grant is still extant and well organized, this claim immediately garnered huge media attention, and soon reporters from *The New York Times*, *Los Angeles Times*, CBS, and NBC were winding their way up the precipitous canyon highway to picturesque Truchas (where John Nichol's *The Milagro Beanfield War* was made into a movie by Robert Redford) to interview Max Córdova, president of the land grant, and the various villagers who had run out of firewood. It made a compelling story— the last vestiges of a rural life being eclipsed by the twentieth century, with villagers suffering their final defeat at the hands of capricious environmentalists taking wood out of their stoves and food out of their mouths.

Everyone had a different idea as to what had caused the firewood shortage or if there even *was* a firewood shortage. William deBuys, historian and author of *River of Traps*, wrote an editorial saying, "The Arizona decision did not create the shortage of dead and down fuelwood in the Truchas area, but it certainly intensified it and brought it to the front page."[45] Sam Hitt of Forest Guardians declared, "It's unbelievable [around Truchas]. You go into the forest and it's like Mexico or third world countries in Asia. There's not a stick left. It's creating huge problems for the wildlife. All 40 species of resident songbirds are in decline."[46] Both agreed that there was a more severe problem in Truchas because it sits right at the edge of the national forest, above the high desert of the Española Valley, and sees more wood gatherers from valley residents in Córdova, Chimayó, and Española than do the villages higher in the mountains. The previous fall the Forest Service had opened up a greenwood area for firewood gathering just down the road from Truchas, but designated a fairly limited number of cords that could be cut. Forest Guardians and Forest Conservation Council responded to the Truchas community by collecting money from donations and buying cords of wood from the Chama area that were hauled to Truchas and cut and split on site.

District Ranger Crockett Dumas's management philosophy was put to the test during the firewood crisis. Urged by Max Córdova and other community activists to respond to the communities' needs and allow greenwood fuel cutting in their traditional social influence zones, on his own authority Dumas sent out his staff to mark trees for a firewood harvest.

Max Córdova viewed both the media circus and Forest Guardians' some-time support with a jaded eye. Surrounded on all sides by both the Carson and Santa Fe forests, Truchas is at the mercy of two different fuelwood policies set by the two different administrations. According to Córdova, both forests had failed to supply enough greenwood thinning areas necessary to meet the fuelwood needs of the community, and in particular, the Santa Fe National Forest had failed to respond to a request from both the land grant and Forest Guardians that the Borrego Mesa area, historically used by Truchas and neighboring communities, supply the necessary 800 cords of wood needed by the communities. In fact, the 6,000 acre Borrego Mesa is claimed by the Truchas Land Grant in a boundary dispute with the Forest Service, and both parties have spent years investigating patents and surveys to determine ownership. The land grant charged that the Santa Fe forest had offered fuelwood in an inaccessible part of Borrego Mesa and that its historical management of the area had favored commercial clearcutting rather than selective cutting. Like Ike DeVargas, Córdova had consistently called for implementation of the Region Three Policy, which mandates that the Forest Service meet the social and economic needs of the indigenous Hispano and Native American cultures.

Max Córdova, right, at Borrego Mesa. Photograph by Eric Shultz.

Max Córdova

Max has always been extremely generous to my family. The other day he asked me if I needed any firewood, and when I said I could always use another cord he said, "Kay, don't ever worry about having enough. Just call me." As an advocate for the land grant, Max was caught squarely in the middle of the battles between the Forest Service, environmentalists, and his community. While the fighting obviously took a toll, it didn't affect the way Max did business: fairly, openly, patiently. Maybe it's because he was trained as a weaver, an art that embodies these qualities. He, his wife Lillian, and two of their children, David and Max Jr., were all weavers of blankets, vests and coats in the traditional northern New Mexico style, which they sold out of a cooperative gallery on the Truchas llano. Business suffered because of Max's involvement in land grant issues, but he didn't complain. He also didn't complain—only lost sleep—over an internal conflict within the grant brought by former officers who relinquished their hold over land grant affairs unwillingly.

Despite all the ranting and raving, there was obviously some common ground that needed to bind both the environmentalists and community activists in a united front. While many accusations of "racism" were hurled about in the heat of the argument—the ramifications of which I'll discuss later in the book—I think the critical issue was the failure to make the distinction between corporate greed and sustainable, locally based resource needs. Both Berde and DeVargas repeatedly called for getting rid of Duke City (a buy-out changed ownership of Duke City from Hansen Industries, a multi-national corporation, to Idaho Timber Corporation). The lawsuit filed by La Compañía and the Vallecitos Association was trying to do just that. DeVargas has said all along that if Duke City were out of the equation, if La Compañía had the funds to buy the sawmill in Vallecitos, and a consortium of small loggers could buy the mill in Española, much less timber would be taken out of the Vallecitos Sustained Yield Unit and the forest as a whole.

Berde and other environmentalists agreed that the Forest Service should start offering smaller sales that would benefit the small operators, but maintained a hard line about the cutting of any amount of old growth—i.e., certain diameter trees—within the Sustained Yield Unit or anywhere on the Carson. Berde insisted that the environmentalists and DeVargas shouldn't be arguing about the La Manga timber sale, that there were other potential sales on the Unit that could provide enough sawtimber from smaller dimension trees than in La Manga. DeVargas, however, said that any sale they might come up would result in the same conflict—there are large trees in all of the areas except those that have been timbered within the last thirty years, and these are not yet ready for reentry. It was imperative that these two groups sit down together because La Compañía settled its lawsuit with the Forest Service, which guaranteed 75 percent of the sawtimber from the 2.1 million board feet in the La Manga sale. The settlement also guaranteed La Compañía 80 percent of the sawtimber from the Agua/Caballos sale, without competitive bidding. But over the course of the next year, as a negotiated settlement of the environmentalists' lawsuit failed time and time again, it became obvious that there would be no substantive dialogue between environmental groups like Forest Guardians and *norteños*. As the ensuing chapters reveal, there would be only rhetoric, threats, and tragedy.

Who *Does* Speak for the Environmental Community?

"History will judge greens by whether they stand with the world's poor."

—Tom Athanasiou

The hard line of environmentalists such as Sam Hitt and Joanie Berde on the La Manga timber sale caused a wide rift in the New Mexico environmental community. The size of the rift was demonstrated at an August, 1996 gathering at the Juan de Oñate Visitors Center just north of Española, where environmentalists, including Green Party members, and *norteños*, including Ike DeVargas, Maria Varela of Ganados del Valle, and civil rights attorney Richard Rosenstock who represented La Compañia Ocho, met to analyze this "bad dynamic." The gathering was organized by other activists who were eager to demonstrate that Hitt and John Talberth of Forest Conservation Council and Berde did not speak for the environmental community at large, and that there are many who support DeVargas and other community activists in their efforts to sustain their communities.

Several themes emerged during the afternoon's discussion. One, forthrightly stated by Chellis Glendinning, a writer and longtime activist from Chimayó and one of the meeting's organizers, was that we are living in an imperialist society where we are treating our Chicano neighbors as our fathers did and their fathers before them. By not supporting land grant claims and the integrity of the Vallecitos Federal Sustained Yield Unit, set aside to benefit local communities, we are essentially opening the door to subsumption by corporate interests and the suburbanization of rural lands. According to Glendinning, the bad dynamic that had been occurring between *norteños* and environmentalists is a result of four behavior patterns: 1) a need to control; 2) a life lived too fast, where the quality of attention is lost; 3) blaming the recipients of colonization instead of the perpetrators; and 4) deifying conflict, where "in your face" politics is promoted as the only way to accomplish anything. To break down these patterns, she said, "we of the dominant culture need to listen and learn."[47]

Ike DeVargas gave a brief synopsis of local history, describing how Hispanos went from being subjects of Spain to that of the United States, with a brief interlude under Mexican control. He explained that his isolated culture is now caught between corporations that want to rape the land, corporations that want to take the land for recreation, and government agencies "that treat us like rats in a lab." He also pointed out that because of the increasing loss of their land base—"in the old days no one ever would have built a house below an *acequia*"—his people have become increasingly dependent upon timber resources.[48]

Maria Varela added that while the businesses spawned by Los Ganados have been largely successful, the people of the Tierra Amarilla area also suffer the effects of a shrinking land base. She warned that Forest Guardians was prepared to file a lawsuit enjoining grazing on the forests of northern New Mexico. She stated: "While none of us has all the answers, some of the traditions that were here are based on sustainability learned over thousands of years. There is a role grazing can play in the environmental balance."[49]

Santiago Juarez, a community organizer who has worked with both DeVargas and Varela, challenged the Green Party members who were present that by not advocating a return of former common lands to communities they are accepting the imperialist idea that these are public lands: "You've given up my sovereignty already."[50] Mike Castro, a Green Party member from Santa Fe, pointed out to Juarez that the Green Party platform did call for the return of some of the common lands, and that they were prepared to support groups like La Companía "if the bottom line ecology is not compromised."[51] He also added that the use of an indicator species, such as the Mexican spotted owl, might not be the best way to do that.

This became a recurrent theme throughout the meeting. Several people pointed out that it is the misuse of the Endangered Species Act (ESA) that is causing hardship in northern New Mexico by limiting grazing and timbering. This point had been brought home by the injunction that shut down all commercial logging in Region Three, without discriminating between large, corporate abuse and small, community-based operations. DeVargas pointed out that the Vallecitos Federal Sustained Yield Unit is on the fringe of owl habitat and the "abuse of discretion" by the Forest Service is the problem.

Santiago Juárez.
Photograph by Eric Shultz.

Sam Hitt and John Talberth were not invited to the gathering, but their names frequently came up in discussion. Varela accused both of "smashing" Madera Forest Products by refusing to release sales it needed to operate. If environmentalists in the room didn't want to be "tarred with the broad brush" of Hitt and Talberth's actions, she said, they should stop the spread of misinformation by being more vocal: "I get suspicious of these democracies where urban environmentalists over speak and out vote us. Until you walk a mile in our moccasins, don't preach to us about what is sustainable."[52] Juarez added to that sentiment when he said that the people in Vallecitos fought for years against Duke City and against clearcuts before Forest Guardians were ever there. "Local communities are not dupes."[53]

In a subsequent meeting held at the Oñate Center to continue the dialogue, DeVargas clarified that while he was encouraged by the support of the environmentalists attending these meetings, nothing could be accomplished unless the Forest Service is made accountable. La Companía just that week won the La Manga decision in district court; the judge ordered the Forest Service to release the sale, but the agency immediately claimed that the logging injunction imposed in the Mexican spotted owl lawsuit prevented them from doing so. Max Córdova, President of the Truchas Land Grant, was also at this meeting, and revealed that his people were currently embroiled in a controversy with the Santa Fe National Forest over the release of two thinning areas near Truchas as commercial sales rather than for personal firewood use. According to Córdova, Santa Fe officials had originally agreed that these thinning areas would be set aside to help meet the firewood needs of Truchas area people (the environmentalists involved in the logging injunction signed off on this agreement), and later reneged on this decision.

The discussion centered on how the environmentalists could help both the loggers in Vallecitos and the people of the Truchas area force the Forest Service to respond to their needs. Tony Povilitis, of the Greater San Juan Coalition in Colorado, suggested inviting the Carson National Forest to cosponsor, along with his group, a workshop to set up partnership projects in northern New Mexico to improve forest conditions and enhance economic benefits. He cited a project in the San Juan Forest of southern Colorado where the Forest Service and community members were already doing this, and suggested representatives from that project be invited to participate in the workshop as well. Everyone agreed that this was a good long-term goal, but also felt that they needed to show their support of Truchas Land Grant members and Vallecitos loggers in a more immediate way. A week later, after several more meetings with representatives of the Truchas Land Grant and DeVargas, it was decided that community members would go out to the controversial thinning areas and cut firewood, whether the Forest Service was there to issue permits or not.

On October 30, Max Córdova, Ike DeVargas, and several Green Party members showed up at the Española Ranger District office with a letter stating that if the Forest Service didn't meet community members at the thinning area on Borrego Mesa and issue the firewood permits, they would harvest the wood without Forest Service permission. On Halloween morning, the *norteños*, armed with chainsaws, headed out to Borrego Mesa. The Forest Service was already there, at the junction of the two forest roads that lead to the thinning areas, with a large, hand-lettered sign reading: "Firewood Permits

Here." A timber staff technician was already at the thinning area, felling trees, which were being sold as "dead and down" at $15 for two cords. At least five armed Forest Service law enforcement officers watched the proceedings (in a later conversation, Max Córdova told me how offended he was offended by their presence). Community members from Truchas, Chimayó, and Rio Chiquito immediately purchased permits and headed up the snow-covered road to the thinning area. Ike DeVargas, Max Córdova, Santiago Juarez, Sam Hitt, various Green Party members, and lots of reporters and photographers headed up with them, and those with chainsaws started bucking up the trees while everyone else helped load the trucks.

In Córdova's letter to the Forest Service he charged that the overall management plan of the Santa Fe Forest is "causing irreparable harm to our communities and our culture." He pointed out that the Forest Service was actually harvesting products from lands patented to the Truchas Land Grant, and that the agency had for years been utilizing clearcutting techniques rather than selective cutting, favoring commercial use over traditional use, which "displaces and destroys culture."[54] The Truchas Land Grant began negotiating with the Forest Service on a firewood management plan that could be implemented in the entire Region Three forest area. They asked that the plan include district advisory boards, implementation of the Region Three Policy, and language that would explain the importance of firewood gathering to the communal traditions of northern New Mexico. Talk of less dependency upon firewood as a fuel source is meaningless unless families can get the money to install solar systems, add greenhouses, or buy natural gas. Once again, the powers that be ignore the fact that activities like firewood gathering continue to connect people to the land. According to Córdova, once that connection is lost, there will be less regard for and less stewardship of the land. "Reckless disregard" of community needs by the Forest Service or professional environmentalists, who favor litigation, will only result in further degradation of national resources. While this time Sam Hitt gave his support to the *norteños*, he acknowledged that there remained serious differences between Forest Guardians and the loggers in the Vallecitos Federal Sustained Yield Unit, and that the La Manage timber sale, as well as future sales, would in all probability only be resolved in court.

In February of 1997, more than 80 people, including many well-known New Mexicans including Lucy Lippard, and Henry Carey signed a paid advertisement (drafted by Chellis Glendinning, myself, and a core group of activists) that appeared in two local papers, stating their support of the rights of indigenous peoples to "pursue the use of their traditional lands so that their unique cultures may survive." The signers made the distinction between wilderness that "should be left alone," and "inhabited wilderness, where people may live safely and sustainably," and perhaps most importantly, be sanctuaries against the encroachment of corporate globalization, an invaluable environmental strategy that everyone could support.[55] And in April of 1997, the Board of Directors of the Southwest Forest Alliance, a group of 55 conservation groups (of which Sam Hitt was a founding member) threw Hitt off the board. The board had met to set future policy, and apparently Hitt stormed out of the meeting after refusing to agree to the board decision not to endorse a position calling for an end to all commercial logging on national forest land. Hitt immediately sent out a press release stating that he had been ousted because of this disagreement. Board members, however, while acknowledging that they disagreed with Hitt on this policy, said he was kicked off the board because of his confrontational style and inability to get along with other members of the group. They also took him to task for his tendency to challenge too many Forest Service logging and controlled burn projects and his lack of commitment to local communities. Kieran Suckling, a board member and director of the Arizona-based Center for Biological Diversity, stated that Hitt's standing in the Southwestern environmental community began to slip in the fall of 1995 when he pushed for the restrictions on firewood gathering in northern New Mexico to protect the owl. Suckling said that most environmentalists didn't think the restrictions were necessary to protect the owl, but yielded to Hitt "because northern New Mexico was his backyard."[56]

In September of 1997 the Ninth Circuit Court of Appeals in San Francisco denied environmentalists' last emergency motion to stop the La Manga timber sale, and on a warm, fall afternoon Ike DeVargas went out to the sale and started felling the ponderosa pines that had become a symbol of everything wrong in a movement that had alienated both indigenous communities and its own community as well. Sam Hitt said to reporters, "We lost. We've been failed by politics to protect the public interest."[57] From DeVargas's point of view, the public interest that Hitt always says he is protecting is the interest of urban environmentalists whose sense of what constitutes environmental health and biodiversity is based on the wilderness ethic of "visitor only." The concept of "inhabited wilderness", where community people live and work, is alien to their experience as weekend recreationists and how they perceive the deep ecology ethic: an environment that separates people from the natural landscape. After La Manga was released, several environmentalists from a reputable Santa Fe organization went out to tour the sale to see what DeVargas and La Compañía had cut: According to one observor, "La Manga is a good prescription."[58]

DeVargas continued to be hopeful that future meetings with environmentalists would prove fruitful, that together they could sit down and actually write timber sale plans that the Forest Service would implement and professional environmentalists would recognize as both protective of communities and resources. After so many million meetings, group affiliations, alliances, and subsequent breakdowns, it seemed amazing to me that DeVargas maintained a faith in people and a faith in the process. In one of our many phone conversations he said to me: "I don't see that the fight is over yet. But we need to leave some of the animosity behind and learn from this experience. The main thing we need to do is educate the people, especially the new people who are coming into our area, as to how we can sustainably live the way we traditionally have. I don't know how easy it's going to be to heal the wounds that were created by this fight. There's a lot of anger out there.... There may be people like Forest Guardians who we can never reach. But we have to reach the other people."

La Jicarita News

"I still view the Forest Service as an army of occupation, occupying northern New Mexico with economic and political force rather than with guns. I'm harsher with them than I am with Sam [Hitt]. But he can no longer portray himself as David fighting Goliath, out to save the poor people against the corporate giant. He's now become Goliath."

—Ike DeVargas

Writing and editing a community newspaper has been an education in itself. While I have always been intrigued with the idea of having a vehicle to say whatever I wanted to say whenever I wanted to say it, *La Jicarita News* has not exactly been that. For several years the paper received its funding under the auspices of the Rio Pueblo/Rio Embudo Watershed Protection Coalition, a diverse group comprised of Picuris Pueblo, *acequia* associations, and environmental groups. While Mark and I functioned as editor and publisher of *La Jicarita*, and every issue disclaimed that the editorials did not necessarily reflect the opinions of coalition members, we had limits. But we still managed to step on any number of toes. After a year's publication, Joanie Berde, a member of the coalition as Carson Forest Watch, took it upon herself to write to the foundations that had supported us with grants—or at least the ones she knew about—to complain that our editorial policy was "hateful" and unfair to Sam Hitt and Forest Guardians for their stand on the La Manga timber sale. The Turner Foundation eventually refused to fund us. (In 1999, *La Jicarita* acquired its own non-profit status, apart from the Rio Pueblo Rio Embudo Watershed Coalition.)

Admittedly, we didn't shy away from that controversy: in the very first issue of the paper, in January of 1996, we interviewed Ike DeVargas about the demonstration he organized in Santa Fe where Sam Hitt and John Talberth were hung in effigy. It was a great interview. He acknowledged right off the bat that by taking an extreme position on the logging and firewood issues the environmentalists were setting themselves up: "I told them every time you mess with the Forest Service, trying to get something good, they'll turn it on you. They turned this firewood thing on them. I told them a long time ago, you guys are too damn extreme...what you're going to do is cause a backlash against the environmental community as a whole. And that would be a disaster. And that is exactly what's happening. They should have made sure that the local communities, the people most affected by the decision, were there at the table when they negotiated the settlement of the spotted owl lawsuit."

We also asked Ike about the fact that representatives of Duke City lumber attended the Santa Fe demonstration. His answer was typical of the DeVargas blunt style: "You know, it's a pain in the butt when you have to end up making a pact with the devil to survive. I hate it. But for me it's survival. If I have to go to bed with the devil to survive, then I'm going to do it. They're not alliances for life. They're alliances on one issue."[59]

Then we went to Joanie Berde for the environmentalists' response to Ike. Berde did express that some strategic mistakes had been made, but maintained an essentially paternalistic attitude about what the environmentalists had done: "Strategy wise, in retrospect, I think that the mistake that was made by the environmentalists participating in the lawsuit was not going to the local communities affected and at least taking the time to explain the nature of the lawsuit and what the impacts of it would be." She claimed that the environmentalists had "learned a big lesson," but any semblance of dialogue or compromise would soon be forever lost, as Forest Guardians and Carson Forest Watch continued their fight against La Compañía in the La Manga lawsuit.[60]

In August, La Jicarita found itself literally in the middle of it. Forest Guardians organized a campout at the site of La Manga Jo—the first of the three La Manga sales—to show the public the sale and to sponsor workshops on treesitting, nonviolence, mountain biking, and mushroom and bird identification. While Hitt said that local loggers were also invited to the campout, the norteños chose to stage their own gathering, and on Friday, August 2, hung Sam Hitt and John Talberth in effigy from trees along the forest road leading to the sale. According to DeVargas, the locals felt that the environmentalists' gathering was "provocative" and counter productive, as all the parties were still involved in legal negotiations over the La Manga sale.

Effigy of John Talberth.

I visited the gathering on Saturday afternoon to interview the concerned parties. The dummies still hung from the trees, accompanied by numerous signs along the route: "It's not the owl, stupid, it's your way of life and culture that's at stake;" and "Enviro-maniacs are the dregs of the 1960–70s hippie movement...Go Home!" The *norteños* were camped alongside the road about a mile before the environmental sentries who manned the entrance to their gathering. People entering the environmentalists' site had to identify themselves and were then given directions along the bumpy, rained-slicked road into the campsite.

When I arrived, small groups were already out hiking several of the timber sale cutting units. Approximately 60 people were in attendance, including an activist group from San Luis, Colorado, called *La Sierra*. Their representative, Praxedis Ortega, a native rancher and farmer, has been involved in that community's struggle to regain access to their lands in the former Sangre de Cristo Land Grant, then owned by the Taylor family (descendants of U.S President Zachary Taylor) of North Carolina. The area was being logged by Southwest Forest Industries, and according to Ortega, the logging activity was threatening the health of the watershed. Ortega and his friends from Colorado attended the New Mexico gathering because "we have to rescue the last of the ancient forests."[61]

I met up with Sam Hitt on a walk through the cutting unit. Hitt denied that the gathering was meant to be "provocative." When asked if there was any room for compromise with La Companía towards resolving the lawsuit, Hitt responded: "This stand of forest must be protected" and that La Companía's insistence on cutting a certain percentage of the large ponderosa pines in the sale was "culturally irresponsible."

I asked if this might be an especially good time to make a concerted effort to form a coalition with the loggers, as the Forest Service was threatening to reduce the number of board feet that will be available to La Companía, and a united effort on the part of the environmentalists and loggers could result in a better management plan for future sales. Hitt responded that while he thought coalitions were important, in this instance there was too much "cultural antagonism," too many differences in the levels of knowledge between the two groups, too much media overplay, and too much Forest Service involvement.

I suggested that perhaps he and his groups were the ones being "culturally irresponsible," that the small, rural communities of northern New Mexico are the last bastion of defense against the suburbanization and urbanization of the area, and environmentalists and *norteños* should work together to ensure that these communities remain economically and culturally viable. While Hitt admitted that this was indeed a threat, he stated his hope that land use plans, like the one being developed in Rio Arriba County, would be able to halt the threat of development. "My bottom line is that these old growth pines will not be cut."[62]

On my way out to visit the *norteños'* gathering, Hitt met me at the security site and made an offer: if DeVargas and the others would agree to take down the effigies, he and some of the environmentalists would be willing to visit their campsite to discuss the issues informally. I delivered the message to DeVargas, who immediately agreed to take down the effigies and said any of the environmentalists who wanted to come would be welcome. I then turned around and drove back to the environmentalists' security site and delivered the message to Hitt. An hour later about eight of the environmentalists arrived at the *norteño* camp, where Hitt, DeVargas, and some of the others, including Max Córdova, sat in camp chairs and talked for about 45 minutes. While there was a lot of finger pointing, name-calling and not much agreement on anything, once the more formal conversation between the two camps was over, Hitt and some of his fellow environmentalists stayed around and engaged in further conversations with the rest of the *norteños*.

When I later asked Ike if he thought the meeting had been worthwhile, he responded: "Actually, I thought it was productive. While we're only 16 days away from a settlement conference on the La Manga lawsuit, and it's too late to do anything on an informal level about that, I think it set the stage for some things that will come up in the future. And I think it was important that Sam talked to some of the others here, not just me, a wild-eyed crazy activist, to see what even a two-month layoff means to working people. We know we have to diversify beyond just saw timber, but until we have the money to buy things like bent lamination equipment—let Sam put his money where his mouth is and buy it for us—we need to go to work."

Entertainment at the community campout.

As he had stated in previous interviews, he stressed that this whole break down could have been avoided if the environmentalists had agreed to release the La Manga sale, except for the 500 acres of old growth, and negotiate with the loggers on how to manage that, or trade it for another potential old growth area. As for Hitt's bottom line that no large trees should be cut, period, DeVargas responded it was likely that very few of the trees that Hitt pointed out to us would actually be harvested.

He reiterated that while it's too late for any negotiations on the La Manga sale outside the courtroom, he thought it was still possible for these groups to form a coalition to discuss future issues: "Let's get together over a period of time so that we can work together to pressure the Forest Service. I still view the Forest Service as an army of occupation, occupying northern New Mexico with economic and political force rather than with guns. I'm harsher with them than I am with Sam. But he can no longer portray himself as David fighting Goliath, out to save the poor people against the corporate giant. He's now become Goliath."[63]

I wrote an editorial in the paper basically taking the Forest Service to task as the entity responsible for letting the situation deteriorate to the extent that every action it takes ends up in litigation. I called for a more open process whereby involved parties meet at the table before projects are released, to insure real and substantive citizen involvement. But over the course of the next few months, the focus of a continuing series of editorials and articles dealt with the deteriorating relationship between communities and environmentalists like Forest Guardians. In the September issue, after attending the meeting of environmentalists and *norteños* at the Oñate Center, we wrote an editorial decrying all the litigation brought by Forest Guardians that had shut down the forests to community groups and pointed out that if Earth First!ers in California could find some common ground between environmentalists and loggers, we could, too. In the October issue we wrote about the other meeting of environmentalists and *norteños* to talk about how to support both the Truchas community in its battles with the Santa Fe National Forest and the Vallecitos area loggers in their battles. The November issue included articles detailing the confrontation on Borrego Mesa and an interview with Max Córdova that expressed his disappointment at a series of negotiations with Sam Hitt and John Talberth where it became clear to him that it wasn't simply a matter of educating them to a different point of view, it was basically a lack of sensitivity.

In December I wrote an article entitled "La Manga Lawsuit Finally Settled." The article stated that while U. S. District Court had ordered the Forest Service to release the timber sale in September, the Forest Service had refused the order because the Region Three logging injunction was still in place. When the injunction was lifted, on December 4, last minute agreements were hammered out among the lawsuit litigants: the environmental groups Carson Forest Watch and Forest Guardians; the Forest Service; and La Companía Ocho, the intervener. The Forest Service agreed to delay the award date of the sale to La Companía until the spring of 1997 so that the logging company would not have to come up with sale money until it could actually get into the unit (too much snow prohibits winter entry). The environmental groups agreed to try and raise $500,000 for a timber buyout: the money would be paid to La Companía in return for their not logging the large diameter trees in the sale, classified as old growth. This compromise was first proposed by former New Mexico Congressman Bill Richardson, whereby the Forest Service would have to come up with $100,000, the environmentalists between $200,000 and $300,000, and the balance from sources sought by the congressman. The Forest Service refused to follow through with the agreement, however, stating that it would be illegal to appropriate money to the loggers, despite the fact that the agency had agreed to compensate loggers in the Northwest to the tune of hundreds of millions of dollars. Richardson also bailed on the proposal, leaving his New Mexico congressional seat to serve in the Clinton administration.

I also stated in the article that because of additional lawsuits filed, it was likely that the La Manga lawsuit would end up in an appeals court, despite the fact that U. S. District Court Judge Mechem declared that all decisions affecting the sale would be heard in his court. This indeed happened, and all the buyout negotiations eventually fizzled out.

Soon after the paper came out Sam Hitt called me up on the phone and told me nothing had been "settled," I was a mouthpiece for Ike DeVargas, and that I was "sick."[64] I invited him to submit a letter and we would publish it, but first we ran an editorial that I'm sure pissed him off even more. The editorial attempted to analyze the internecine struggles within the environmental community, both locally and nationally: "Those activists who come to the environmental movement with a background in social justice issues—labor organizing, civil rights, the New Left—are often called social ecologists: They see human beings as an integral part of the natural world that is being manipulated and exploited by the industrial, capitalistic economic system. Other activists, often called deep ecologists, come to the movement to save our wildlands as a moral statement apart from any value these lands may have to human culture. In the introduction to a book entitled *Defending the Earth*, in which social ecologist Murray Bookchin and Dave Foreman, one of the founders of Earth First!, come together to try to find common ground between these two philosophies, editor Steve Chase says: 'While social ecologists...trace the roots of the ecological crisis to the rise of hierarchical and exploitative human societies, many deep ecology activists talk of the human species itself as a blight upon the planet.... Indeed, the deep ecology movement as a whole lacks a consistent or clear social analysis of the ecology crisis or even a consistent commitment to humane social ethics.' "[65]

Hitt's letter, published in the next issue, called into question the concept of "sustainable sanctuaries"[66] because, in his opinion, these communities are being excepted from environmental safeguards, and that the survival of an ecosystem-based culture rests on laws that protect that land. He of course never addressed the issue that these laws are Anglo-driven, colonialist, often used indiscriminately, and are sometimes in need of fine-tuning. Whoever dares to initiate this dialogue has been labeled "Wise Use", a term that refers to a politically reactionary movement in the west that advocates a balance between environmental protection and economic need but has essentially been a smokescreen for a corporate attack on environmental laws.

At a meeting of the Santa Fe Group of the Sierra Club, whose members have refused to support Forest Guardians' "Zero Cut" (no commercial logging on public land—see Zero Cut chapter) and the national Sierra Club's "No Logging on Public Lands" initiative, members of Forest Guardians (who were also Sierra Club members) showed up to disrupt the meeting and proclaim that *La Jicarita* was indeed a Wise Use publication. They also called Santa Fe Group members Wise Use, particularly the three members Courtney White, Barbara Johnson, and Jim Winder who founded a new organization called The Quivira Coalition, which hoped to establish dialogue and trust between ranchers and the environmental community to restore and maintain both healthy ranches and healthy ecosystems. They work with a number of government agencies and sponsor workshops and tours that promote good resource ranchland management. As White wrote in the first coalition newsletter: "Coalition members believe that ranchers and environmentalists share too much in common to keep fighting: love of land, for example, and a desire to experience solitude and beauty. We believe that open space is vital to the future of this country and its quality of life. Since ranchers play a critical role in the protection of our wide- open skies, we should be allies fighting the spread and stink of urban sprawl, which threatens both the health of the ecosystem and the vitality of ranching. The future of the West depends upon our ability to shake hands and get to work preserving open space."[67]

Hitt and other absolutist environmentalists continued to attack *La Jicarita News* over our coverage of their appeals and lawsuits; it reached its nadir in 1999. On March 19, a pipe bomb was found in the mailbox of Forest Guardians in Santa Fe. Fortunately, the bomb failed to go off, but several days later the group received an envelope in the mail with a drawing of a rifle scope's cross hairs over the words "Forest Guardians" and "see-ya" written underneath. It was signed "MM—The Minute Men."

On March 23, Charlotte Talberth, former Forest Guardians board member and wife of Forest Guardians' then-executive director John Talberth (formerly of Forest Conservation Council), sent the following e-mail to a person or persons unknown:

"Today at 10:15 am Forest Guardians' chief canvasser, Mike Cherin, discovered a pipe bomb when he opened the mailbox outside Forest Guardians' office in Santa Fe. The bomb did not explode and no one was hurt. Police bomb squads subsequently arrived and detonated the bomb, which was loaded with ball bearings, meaning it was designed to kill.

Last week Rio Arriba County hosted a day-long meeting of officials from counties in Northern New Mexico and Southern Colorado. The purpose of the meeting was to get more counties to join Rio Arriba in intervening against Forest Guardians et al. in the national economic lawsuit, and to censure the State of the Southern Rockies report authored by John Talberth and Bryan Bird (a Wildlands report).

The people who spoke at the meeting are all known to us—Ike deVargas, Santiago Juarez, Chellis Glendinning and Richard Rosenstock. They foment hatred and violence against Forest Guardinas and Zero Cut on a regular basis by calling us genocidal racists etc. in public forums and publications, particularly La Jicarita, a newsletter circulated in Northern New Mexico. According to Bryan Bird, who was there, numerous threats of violence were made at the meeting.

We are very, very thankful that Mike was not killed. I can post whatever stories are printed in tomorrow's papers. Mike says it makes him want to work harder than ever."

Andy Caffrey of the EarthFirst! Media Center somehow received the e-mail and immediately sent it to a "Recipient List Suppressed." All of us who "foment hatred and violence" soon saw a copy of it. Talberth attempted to do damage control with a subsequent e-mail to Andy Caffrey stating, "had I known that the email was going to be forwarded to you and then broadcast, I would have been clearer in stating that I have no reason to believe that any particular individuals were directly involved in the attack. It was never my intention to add to the already divisive dynamic that exists between people who should by rights be allies: those who claim to speak for local rural people and those who claim to speak for the environment."

No one was ever arrested for placing the pipe bomb in Forest Guardians' mailbox.

The Art of Irrigación

"Here's a land where life is written in water."
—Thomas Hornsby Ferril

O ur ten acres of land are irrigated from the Acequia de Arriba and the Acequia de Abajo. Each *acequia* is diverted from the Rio de las Trampas above the village, where *presas* or dams send the water through the *acequia madre*, or main ditch, to every *parciante's* (irrigator's) *compuerta*, or headgate. Here, turn by turn, we take the water out of the *acequia madre* and deliver it to our fields via laterals (the Spanish names for which vary from area to area).

The ability to guide the water from the laterals over an entire field that undulates with the terrain and follows arbitrary boundaries is an art form, to say the least. In his book *River of Traps, A Village Life* William deBuys tells the wonderful story of how Jacobo Romero, the master artist, attempts to teach Bill how to deliver the water to his field, parts of which never progress beyond what Bill calls "Albuquerque," referring to their desert-like qualities.[68] When it was Jacobo's turn to irrigate, he, of course, headed to his fields to monitor and guide the water, which is what is necessary to insure success. The rest of us, caught up in 20th century madness, have barely the time to open the *compuerta* at 6 in the morning and remember to close it the following morning at the same time (one *derecho*, or water right, in our village means you get to use the water for 24 hours about every two to three weeks, or as long as it takes for everyone to take shares according to his or her right). We have an area of land in the lower field we refer to as the "Sahara."

Acequia politics are even more interesting than the art of irrigating. The irrigation system is governed by an *acequia comisión*, comprised of three *comisionados*: a president, a secretary and a treasurer. These commissioners are elected by the *acequia parciantes*, usually for two-year terms, and they are essentially an executive committee that oversees any work or repairs to the *acequia*, keeps records and account books, and makes all decisions pertaining to *acequia* administration. Day to day management of the *acequia* is the domain of the *mayordomo*, also elected by the *parciantes* or appointed by the commission, to oversee the scheduling of water delivery to each irrigator. Unfortunately, *elected* is sometimes not the right word to describe how *mayordomos* are chosen, in these days of obligatory work that takes many community members out of the villages from eight to five. More often a *parciante* is "persuaded" or "arm-twisted" into serving as *mayordomo* because it takes a lot of time and effort and diplomacy to make sure everyone gets his or her share of water delivered in a timely fashion.

El Valle ditch cleaning crew.

This was not so in our village when we arrived. We fought over who got to be *comisionado* or *mayordomo*. We were comprised of a generation of older men who were born and raised in the village and who often had to leave to find work or careers in other cities or states, but who all came back to retire. While they were all willing to serve the *acequia*, I must point out that they were partly motivated to do so because they didn't want certain other people to be the *mayordomo* or *comisionado*. Fortunately, even for relative newcomers like Mark and me, it was easy enough to keep track of just who was who's enemy: you were either allied with Alpha Male Number One, Tomás, our next door neighbor, or you are allied with Alpha Male Number Two, whom I'll call Alfredo. There was never any question as to where our allegiance lay—with Tomás—although we made every effort to not overtly take sides, at least in our day-to-day relations. It was difficult, however, to maintain independence when you were in the middle of an *acequia* meeting where hands were raised to elect officers and then brought together in vengeful applause when victory was achieved.

How these two sides came to be enemies is a story unto itself. My story has to do with an *acequia* meeting Mark and I attended to elect officers and *mayordomos* of both of our *acequias*. Alfredo had done some homework and came to the meeting prepared for an unfriendly takeover. He and his supporters suddenly pulled out the bylaws governing the *acequias* and pointed out they stated that each *parciante's* vote was determined by the number of *derechos* (water rights) he/she owned. Of course, Alfredo and several of his buddies are big land owners who each own three or four water rights and they immediately said this was how we were going to vote and before Tomás and the rest of us could respond, they'd out-voted us. The results: Alfredo took total control of one of the *acequias* (he was elected both a *comisionado* and the *mayordomo*, which we immediately objected to and were overruled). Unfortunately, we didn't even have enough votes present to challenge the voting method, as some of our *parciantes* had neglected to attend the meeting (many don't come because the meetings are so often contentious). It was one of the worst meetings I'd ever been to—and I've attended a lot of hostile meetings in my day—because these were my friends and *vecinos* arguing over our life's blood.

It's an argument that reached the New Mexico courts. A number of years ago, in La Lama, a community north of Taos, a wealthy Anglo landowner decided he'd do the same thing that Alfredo did and suddenly challenged the voting process on his *acequia* to gain control. The rest of the community joined against him, hired a lawyer, and successfully challenged his position in court. The ultimate decision was that each *acequia* association has the right to determine its own voting practice, either by *derecho* or by *parciante*. This was a victory for the La Lama community, which immediately adopted the more democratic system into their bylaws. So the precedent had been set for us. Now all we had to do was get our side together, make sure they all attended the next *acequia* meeting, and voted to amend the bylaws to state that officers will be elected by one *parciante*, one vote. This was after we figured out how many votes it was going to take to change the bylaws and assure that we voted for the changes by the one *parciante*, one vote method.

Fortunately, while all this rancor was playing itself out, it proceeded to rain so hard all summer that no one was interested in taking any irrigation water, the *mayordomos* had almost nothing to do, and we all went about our business as usual. But of course there's always the next summer when other disputes and policies will be argued out at *acequia* meetings. But we're not unusual; every village in every corner of New Mexico plays its own particular feud out in its own particular way, and we continue to march on to the beat of our different but equal drummer: *el agua, la cultura, la tradición.*

Sipapu

"They're just using all the water they want without the right to use it. As farmers on acequias, we follow the rules for how much water to use and when. What in the world makes them feel they can do what they want with water without following the same rules as the rest of New Mexico?"

—Harvey Frauenglass, Acequia Comisionado

Not only did we jump full speed into the controversies surrounding communities and environmentalists in the first issue of *La Jicarita*, we also wrote an editorial against the proposed Sipapu Ski Area expansion, and became, in the minds of ski area supporters, elitist environmentalists. There was more than a little irony to that accusation: the ski area supporters were largely ski area instructors and folks who skied there, while the ski area detractors were the housekeepers and downstream farmers, including Picuris Pueblo. The ski area instructors especially didn't like our questioning "the legitimacy of using economics to justify expansion when ski areas primarily provide menial, minimum wage, and seasonal employment." Two of them immediately fired off letters to the paper that their jobs were not "menial" and they were offended. My neighbors down the road, however, who had worked at the ski area as housekeepers, didn't object to the "menial" assessment: "I wouldn't work that shit job again even if the Bolanders [the owners] paid me a living wage."

One of the ski instructors who sent a letter was the mother of one of our son Jakob's closest friends. Jakob, along with a handful of other teenage kids, mostly Anglos, all worked at the ski area as "angels", or part-time instructors who got to give a lesson when the ski area had more clients than regular instructors on hand. Mostly, they skied—for free—which is why they were willing to work for $20 a day. The mother's letter set the tone for the ensuing battle: an extremely emotional defense of the Bolander family and their contribution to the community. There was little analysis of what impact an expansion might have upon the community and its environment; no acknowledgement that the Bolanders were illegally using surface water to make snow; and certainly not much sensitivity to Picuris Pueblo concerns that the ski area was right in the middle of its sacred lands and archeological sites, let alone that its name, "Sipapu" was an irreverent use of a religious word. If *La Jicarita* stuck to the business of opposing copper mines or timber sales we were an asset to the community; once we struck too close to home—a locally owned, fun place to ski—we weren't being community minded.

Jakob was caught in the middle, of course, and while I felt a certain amount of guilt I also realized it wasn't as if he were the son of the Rosenbergs or black listed (supposed) communists during McCarthyism. He handled the whole business fairly well, even when the Bolanders threatened to not hire him as an angel because "the bad feeling between your parents and us might make things difficult." He stood up for himself, and I think admired his parents for their commitment, despite the censure of the parents of his peers.

The ski area owners and their supporters deliberately started a campaign to accuse all of those who opposed the expansion of wanting to shut the ski area down. There wasn't even any subtlety to the campaign. They could have said, for instance, that as a result of our opposition the ski area might not be economically viable, or it might be unable to secure enough water to operate its snowmaking machines. No, for reasons that had nothing to do with our concerns about overuse of water or secondary impacts of ski area development, *La Jicarita*, farmers, *acequia* associations, Picuris Pueblo, and all those "environmentalists" wanted to shut Sipapu down. This was our agenda, so they said, despite the fact that many of our kids had participated in the school ski program, some of them worked there, and Mark and I had volunteered with the ski school program for three years.

In many ways, the ski area expansion would been much easier to oppose if it had a corporate owner rather than the Bolanders: Lloyd, his wife Olive, and their son Bruce. Lloyd and Olive, who had grown up in the area, got a forest permit to run their small ski area back in the early 1950s. They didn't even start making snow until the mid 1980s. Before then, when mother nature was stingy with her wintertime snows, there were years when the area didn't open. To supplement the skiing, the Bolanders also began developing the area as a summer resort: many Texans come with their RVs and off-road vehicles (ORVs) to stay in Sipapu motel units or to participate in resort-sponsored ORV activities. Despite Lloyd's Republican Party affiliation and his libertarian approach to running a business—I'm the ski area expert so leave me alone to do business as I see fit—much of the local community respected him as a decent man who basically just loved to ski.

Feelings about his son Bruce, however, who became president of the business in the 1980s, were different. Even when it became obvious that the Bolander's hands-off attitude was not going to work anymore in the '90s, as community activists, federal and state agencies, and Picuris Pueblo began to take a closer look at what exactly was going on over at Sipapu, Bruce was resistant and pugnacious about fighting every attempt to question or regulate water or land use for the proposed expansion.

The Rio Pueblo/Rio Embudo Watershed Protection Coalition appealed the Sipapu expansion Environmental Impact Statement (EIS) in 1995. The EIS was withdrawn when a Forest Service review acknowledged that the service had failed to adequately consult Picuris Pueblo with regard to its cultural and historic properties in the area. Members of the coalition who are downstream *parciantes* protested the ski area's proposed transfer of water use from agricultural to snowmaking, and soon found out that the Office of the State Engineer (OSE), which regulates water use and is the agency responsible for change of use applications, was investigating whether the ski area actually owned any water rights in the first place. The OSE also informed Bruce that his commercial use of a spring on the ski area property was illegal (it was regarded as surface water and therefore he needed a permit to use it for commercial purposes) and delivered a "cease and desist" order.

Bruce ignored the "cease and desist" order (the OSE has little enforcement capability) and refused to participate in any kind of pre-hearing negotiations on his water transfer request. For a while, before any hearing was scheduled with the OSE, he came to coalition meetings thinking that perhaps he could persuade the members to drop their protest altogether and agree to his proposed expansion. When it became obvious to him that unless he was willing to compromise on the size of the expansion and the amount of water he wanted for snowmaking, the *parciantes* were going to follow through with their protest, he quickly became adversarial and started spreading the word that we wanted to shut him down. He also hired two lawyers to represent the ski area at the OSE water transfer hearing. The OSE decided to pursue the validity of the ski area's water rights, and with the help of the *pro bono* lawyer representing the *parciantes*, declared that the Bolander family had forfeited their water rights in the early 1960s (because they had allowed the state highway department to obliterate their *acequia* during highway realignment and had therefore not used water for four consecutive years to irrigate, as state law requires). The ski area therefore had no water to transfer, or use for commercial purposes, and if it wanted to make snow would have to buy or lease water rights.

Bruce largely directed his hostility at *La Jicarita*—Mark and me. He accused us of being fanatical in our opposition to the proposed expansion and the "brains" behind the battle. This attitude, seen as very patronizing by the rest of the coalition, of course, only increased everyone's resolve that the Bolanders comply with water law and scale back the expansion. Thirteen *parciantes* signed a protest of a second water transfer application the Bolanders sought to transfer water from eleven downstream acres owned by Lloyd and Olive on the grounds that the transfer would impair other *acequia* users and was contrary to the public welfare. Picuris Pueblo again raised the issue of the ski area name, and as part of the latest compromise proposal made by the coalition members, asked the ski area to change it. We also asked that they extend the present ski lift vertically (dropping a second, longer lift to the west), drop plans for a second restaurant at the top of the lift, meter all their water use, including snowmaking and water for their motel units and amenities (summer use was a big concern), comply will all state water law, and give precedence to agricultural water use in times of drought conditions.

No one was hopeful that Bruce would agree to any of these conditions, and the second water transfer protest was denied, allowing the Bolanders to transfer downstream water rights to the ski area for making snow. The Forest Service re-released the EIS, which was again appealed. Crockett Dumas, district ranger, initially expressed regret that the district released the document in the first place without more community input, as dictated by the Collaborative Stewardship process, but subsequently bowed to pressure from ski area supporters and released an amended EIS that failed to address any of the concerns regarding water quantity, water quality, or cumulative impacts related to ski area development. Meanwhile, the ski area applied for its fourth emergency permit to make snow for the winter of 1998–1999 and was approved by the OSE. The rumor continued (a rumor that had been around since the Bolanders first applied for an expansion) that they were ready to sell: the only reason Bruce was so unwilling to compromise on any aspect of the expansion was that he wanted all permits in place before the family sold to an outside interest.

While the Bolanders never did implement the expansion proposed in the EIS, they did expand the ski area within its permitted boundaries (another story of Forest Service ineptitude and court injunctions and lawsuits), and in 2001 they sold majority interest in the ski and summer resort to investors.

The Empty Nest, a Year Early

"There's no place like home."

—Dorothy in *The Wizard of Oz*

The only reason I could say I was happy that Jakob, at seventeen, decided to become an exchange student in Spain for his senior year was that it meant he wouldn't be working at Sipapu. Mark and I wanted as little contact as possible with folks whose mission was to defame us.

I'd been warned that I would be depressed as hell when Jakob left. So when he took off for Zaragosa in August of 1998 on a foreign exchange scholarship, I was intellectually prepared for a profound sense of loss while Mark hadn't prepared himself for anything. Emotionally, two weeks after he left, we were both wrecks. It took all our fortitude during his first phone call to refrain from telling him to get on the first plane home.

During the first few months he was there I'm sure Jakob would have been more than willing to do so. Here was a kid who had lived his entire life in a rural community suddenly in a city of 600,000 people (we constantly reminded him to watch out for traffic). Here was a kid who thought he knew a little Spanish—having taken classes in high school and having been around Spanish-speaking people most of his life—who found he didn't understand a word. And here was a kid whose high school classes had consisted of seventeen students who called teachers by their first names who was suddenly sitting in classes of 40 students listening to teachers who didn't even know *his* name.

Despite his courageous attempts at rising to the occasion—"I know I can do this" and "I know I'll be okay"—he was lonely and depressed for weeks. He dropped two academic subjects (he still had a full load of art history, geography, literature, language, and history), he chased down American students on the street so he could speak English, he e-mailed his girlfriend back in Taos twice daily and us at least every other day, and he stuffed himself with bread all afternoon long to make it to the 9:30 dinner hour.

Then, suddenly one day, everything was okay: he passed all his academic exams, in Spanish; he broke up with his Taos girlfriend and acquired a Spanish one; he joined a mountaineering club to go rock climbing and skiing; he went to political rallies for the Zapatistas; he visited the Gaudi museum and saw Dalis and Picassos in Barcelona; and he realized he was having the time of his life.

Mark and I, of course, weren't having the time of *our* lives, but as the days passed the pain subsided or at least settled into a place that made it less acute. There was bitching, too, about all the work *we* had to do that *he* should have been doing to get his college applications off to five schools by February 15: sessions with the high school counselor; running down teachers for recommendations; filling out financial aid forms that reveal your entire financial worth or lack thereof; e-mailing back and forth trying to help figure out what exactly colleges expect from seventeen year-olds' essays about the meaning of life.

The Hanukkah/Christmas time was bad, of course, despite ski holidays on both sides of the ocean. Jakob's nostalgic vision was sitting next to our wood stove in his T-shirt reading a book (apparently his family's large condominium was hard to heat). When we told him that even if he'd been here the fire would have sufficed only so long and the lack of snow at Sipapu (only the beginner area was open for Christmas) would have had him climbing the walls with boredom, he took off for the Pyrenees where he skied and boarded for four days of bliss. We vowed we would all be together for the next Christmas, no matter where he was for college.

The rest of the year went quickly for him, and in May Mark, Max and I showed up for a visit. It was a wild two weeks. We'd tried not to be overly ambitious, but of course we were. How can you not be when this is the only time in your life you may be able to visit a country that looms very large in your cosmology (the connections between Spain and northern New Mexico are complex, to say the least) and where your 17-year old son has just spent a year without you. Jakob and his girlfriend Laura met us at the airport in Madrid, where we spent several days trying to get to know her in our bumbling Spanish and seeing the El Grecos and Velázquezes and Picassos in the Prado and Reina Sofia. Jakob had taken an art history class in high school and took us on a tour of Spanish architectural history throughout the entire visit. Spaniards take their history very seriously and spend many weekends and holidays inspecting their Moorish, gothic, and Renaissance architecture, much like we in the American west visit our own at the ruins of Chaco Canyon and Mesa Verde. Jakob's host mother, whom we visited in Zaragoza, told us that she had very little interest in visiting the United States (the architecture of New York compared to the Alhambra or Barcelona?) although she would like to see places like the Grand Canyon or Teton National Park.

We saw and did a little bit of everything: the olive trees on every square inch of southern Spanish land; the windy streets of Granada in a rented car desperately trying to find a hotel; the magnificent Alhambra and Generalife gardens, where *acequia* culture began; the overbuilt Mediterranean sea coast to Barcelona; the amazing transformation of our son from a country boy to city boy as he led us through the subway maze beneath the streets of Barcelona and Madrid; the obvious affection between him and his family of 10 months; and the monasteries and peaks of the Pyrenees, reminding us of home. The rivers were polluted, but the fields were productive and the food sumptuous. We reconnected with our Spanish-speaking son in bouts of pleasure and frustration, and we all came home to a rainy and productive summer of our own, which saved us from a winter drought. In August he was off again to college, but only as far as the University of New Mexico, where we saw him almost every month and he busily integrated his Castilian Spanish and *norteño* life into a major field of study (he eventually graduated from the University of California at Santa Cruz).

"Zero Cut"

> "In today's corporate global economy, the vanguard of environmentalism becomes not just the conservation of pure wilderness against the thrust of human civilization: it becomes the fostering of human survival through non-racist, non-sexist, economically equitable community living in direct and sustainable relationship to the Earth."
>
> —Urban Habitat mission statement

The "Zero Cut" logging initiative of Forest Guardians engendered both backlash and internecine conflicts within environmental groups that took a stand on the issue. On June 4, 1997, a full-page ad appeared in *The New York Times* and *The Santa Fe New Mexican* entitled "Zero Cut Now." The ad, which called for a ban on all commercial logging on public lands (and solicited funds for Forest Guardians), was signed by a handful of national environmental groups including Earth Island Institute, the Constitution Law Foundation, Protect Our Public Lands, and Rethink Paper (a project of Earth Island Institute). Conspicuously absent was the Sierra Club, which a year before had passed a referendum to end commercial logging on public lands.

When I called Bruce Hamilton of the national Sierra Club conservation staff to ask why the Sierra Club had not signed the ad, he said there had been disagreement over the wording. The Sierra Club felt that calling the ad "Zero Cut" would raise a red flag that might compromise the intent of the initiative, which was to ban commercial industrialized logging on public lands. Using the words "Zero Cut" could give the wrong impression that the Club was opposed to cutting personal firewood or "cutting one's own Christmas tree." He told me that the Sierra Club would run its own ad in *The New York Times* to support Representative Cynthia McKinney's National Forest Protection and Restoration Act, introduced on October 31, 1997, that called for an end to commercial logging on all national forest and other federal public lands. The bill would leave intact firewood collection and "other traditional personal uses of the forest." It also called for a National Heritage Restoration Corps to restore federal forest lands to their natural condition and redirect logging subsidies to provide funds for worker retraining.

Hamilton admitted that even its own more carefully worded initiative had raised controversy within the rank and file of the Sierra Club. While Hamilton stated that members who disagreed with the logging ban "should not be allowed to speak for the Sierra Club," the national organization authorized chapters to formulate their own forest management strategies based on local needs and politics.[69]

The Santa Fe Group of the Sierra Club (later called the Northern Group) is one of the local groups that refused to support the ban. Courtney White, then chair of the Conservation Committee, explicitly explained his reservations regarding the "Zero Cut" policy in his column in the Sierra Club publication *Rio Grande Sierran*, pointing out that if "we kill off rural communities" it is the developers, not "wildlife and other agents of biodiversity" that will step in to fill the vacuum. He chastised the membership that a no-logging policy is "elitist" and destroys the Club's ability to effect change and improvement of activities on public lands.[70]

As a result, David Orr, chair of the national Sierra Club No Logging Task Force, sent out a message on the Club's listserve attacking White and calling for his censure: "I call on the ExCom [Executive Committee] to take steps to bring the *Rio Grande Sierran* in line with national club policy. And I call on the ExCom to issue a formal apology in the next issue. This is an absolute disgrace, and all those who were elected to represent the members of the Rio Grande Chapter [parent of the Santa Fe Group] should be ashamed of yourselves. It's time to stop printing Wise Use rhetoric in Sierra Club publications."[71]

Hamilton told me that he felt Orr had overstepped his bounds in calling for the censure of the Rio Grande Chapter of the Sierra Club, but reiterated that the national organization would continue to push for an end to all commercial logging, even if this policy failed to differentiate between corporate and small, locally based logging.

The Santa Fe Group found itself battling not only a national mandate but for its very survival. In the August 1998 issue of *La Jicarita* White lamented the Club's drift towards confrontation and no compromise on issues like commercial logging and specifically pointed to Forest Guardians' rising influence in the state chapter: "A determined effort by members, family, and friends of Forest Guardians to bend the Chapter to their 'take no prisoners' conservation philosophy is now underway." Several Forest Guardian members held leadership positions in the "Zero Cut Campaign" (which later became the National Forest Protection Campaign) and had been trying to get elected to chair and executive committee positions in the Rio Grande Chapter of the Sierra Club. White went on to say how this was affecting the group as well: "Forest Guardians has been trying to strong-arm the Santa Fe Group of the Sierra Club for the last six months, demonstrating, through their actions, that they will not tolerate dialogue and collaboration. They certainly do not tolerate dissent."[72]

The issue that finally turned theoretical disagreement into all-out war was the Agua/Caballos timber sale in the Vallecitos Sustained Yield Unit of Carson National Forest. As I discussed previously, in 1995 La Companía Ocho was guaranteed 80 percent of the Agua/Caballos sale as settlement of a lawsuit the organization brought to force the Forest Service to comply with the terms of the Sustained Yield Act. The Forest Service released the Agua/Caballos Draft Environmental Impact Statement (DEIS) in 1999, and the Santa Fe Group's forest issues chair, George Grossman, submitted comments to the Forest Service supporting Preferred Alternative C, which called for a harvest of 10.6 million board feet in a 6,400-acre sale area. According to Grossman, the heavily overstocked sale area of the Unit could not only support this kind of a cut but was badly in need of it. In his letter Grossman acknowledged that while Sierra Club policy called for no commercial logging on public lands, at that time there was no legal mandate behind this policy and that it was impossible to manage the forests, under current budget constraints, without commercial logging. His comments were approved by the group's conservation chair, Cliff Larsen.

All hell broke lose two months later when Bryan Bird, the Conservation Biologist/Appeals Coordinator of Forest Guardians, who was also Secretary of the Rio Grande Chapter of the Sierra Club, filed a complaint stating that Grossman "violated club policy, misrepresented the Sierra Club, and misused the Sierra Club name and letterhead in an official capacity." He also stated that Grossman "should step down from his position with the Santa Fe Group and no longer volunteer his services to the Club."[73] Why he waited two months to file his complaint was unclear, but according to Santa Fe Group members, Bird submitted his letter to at least one national board member before he submitted it to the group, which contravenes Club procedure. Accusations among the group, chapter, and national started to fly once Bird's action became public, and the situation quickly degenerated into a Trial by Email, as several group members called it.

The implication of any kind of "Trial" of George Grossman was offensive to many Club members and environmentalists familiar with Grossman's long and illustrious history. In particular, as Carson Forest issues chair for the group, Grossman was intimately familiar with the entire forest, was actively involved in the creation of the Forest Plan in the mid-1980s, and had successfully fought inappropriate timber sales like the Angostura on the Camino Real Ranger District. In fact, he was named a National Sierra Club Environmental Hero in 1992 for his work.

The inherently hierarchical nature of the Club became apparent when the National Conservation Governance Committee issued a letter to members of the executive committees of the group and chapter requesting that a formal letter be sent to the Forest Service withdrawing the support of the Sierra Club for Alternative C, and that any further communication with the Forest Service or the press be made jointly from the group, the chapter, and the national. The group and chapter were given less than a week to address this request.

Many of the local members viewed this letter as essentially a threat and a gag order. That opinion was expressed at a chapter conservation committee meeting on June 19, but the committee voted to bring the national's directive regarding Grossman's letter before a chapter "issues" committee. At the request of this committee, chapter chair Gwen Wardwell wrote a letter to the Forest Service stating that Grossman's comments were being replaced by the chapter's comments submitted in her letter. (The Forest Service stated that Grossman's letter would remain in the official comments to the DEIS.) Wardwell's letter went on to say that because the Sierra Club opposed any commercial logging on public lands the Club could not support any alternative in the DEIS. She then expressed the Club's generic opposition to road building, its concerns about endangered species, forest restoration efforts, etc.

Wardwell's letter insured that the Sierra Club would be absent from the negotiating table where the Forest Service, community members, La Compañía, and environmentalists continued to engage in dialogue about the sale. Although the official comment period was over, the Forest Service had been receptive to additional input regarding Alternative C, and had already modified its analysis and proposed action.

As members of the Rio Pueblo/Rio Embudo Watershed Protection Coalition Mark and I had been involved in sale negotiations and were invited by several Santa Fe Group members to make a presentation at the June 19 conservation committee meeting regarding the timber sale and the ongoing negotiations. The meeting was held at Camp Summerlife, near Peñasco (White and several others thought it appropriate for Sierra Club members to actually come to northern New Mexico if they were dealing with issues that affected northern New Mexico) and Mark and I brought along some *norteño* activists for support. After our presentation, which defended the sale as a good example of a situation where loggers and environmentalists were working together to come up with a good prescription, a member of the El Paso Group complained, "I didn't drive nine hours to listen to this crap." And in what some participants later described as a "racist and offensive diatribe," an Albuquerque member said: "There is a tendency for this group to wander away from the pure environmental focus to the sociological.... When I graduated from high school I left home with $123 in my pocket and I educated myself.... I would love to live in a small rural community. Every day I go to work and walk 300 yards down a hall with no natural light to a room with no natural light to work so that I can make sure that my family is fed and I can educate my children [he identified his work place as Sandia National Laboratory]. I have had to choose to nail myself to this particular cross. If I live long enough I hope to enjoy life in some of these small villages, if they are still there.... People seem to think they have a right to live rurally and they can take it off the backs of the taxpayers any way they want."[74]

Several weeks after Wardwell's letter was sent to the Forest Service an article in the *Albuquerque Journal* addressed the fact that environmental groups are overwhelmingly white and exclusive. Bruce Hamilton (of the Sierra Club's national conservation staff) was quoted as saying that there was no denying that the Sierra Club is predominantly white and they needed to strive for the diversity of membership and staff. But what the *Journal* article failed to address, even with this talk of minority exclusion, is exactly what effect these policies have on people's lives. A Sierra Club policy voted on by an overwhelmingly white, middle-class constituency to oppose commercial logging on public lands (less than 10 percent of that overwhelmingly white, middle-class membership bothered to vote) is an example of reductionist thinking that separates facts and values and fails to differentiate between sustainable economics and corporate greed, fails to recognize the importance of land-based work and culture, and homogenizes diversity. Other nationally directed policies—or attempts to set policy, like the immigration ban policy, which failed to pass (see The Sierra Club: Measure A or Measure B Chapter)—effect the same kinds of discrimination in other parts of the country.

During this time of heightened tension between *norteños* and environmentalists, a group of us had been meeting informally to support each other's work on issues of community forestry, water rights, land grant restitution, and economic justice: Chellis Glendinning, Ike DeVargas, Max Córdova, Eric Shultz, Mark, and I were already immersed in forest and land grant politics while David Benavides and Pat D'Andrea primarily worked on water issues, David as a water rights attorney (with Northern New Mexico Legal Services) and Pat as liaison between the environmental and social justice communities. Lisa Krooth, as director of Legal Services, provided agency support for La Compañía's struggles to remain extant and for David's work for *acequia* communities. We invited George Grossman to participate in "El Grupo," as we humorously referred to ourselves, because of his battles within the Sierra Club over the "Zero Cut" policy and as a representative of the Santa Fe Group, whose members recognized that environmental issues do not exist in a social vacuum. Jake Kosek, doing field work in Truchas for his PhD dissertation and a good friend of Max Córdova, Mark, and me, was an honorary member (Kosek later wrote a definitive book about our struggles: *Understories: The Political Life of Forests in Northern New Mexico.*)

As the pressure intensified on the Santa Fe Group we decided upon a strategy to bring attention to the hierarchical nature of the Sierra Club and the absolutist policies—restricted immigration, "Zero Cut" and "Zero Cows" on public lands—that discriminate against and oppress rural and urban minorities. A few of us were already members of the Sierra Club; several more of us joined, and we decided to call a press conference to announce our resignation from the Club and the formation of our new organization, El Grupo. It was purely a publicity stunt, as none of us other than George were active members of the Club and we'd already been meeting as El Grupo for quite awhile, but we figured it would help bring the debate before the public eye and shake up some folks in the Sierra Club.

We sent out a press release stating that the Sierra Club state chapter had placed a gag order on George and the Santa Fe Group, quoting Ike DeVargas's response: "I am concerned about the heavy-handed tactics employed by the major environmental groups such as the Sierra Club to stifle dissent within their own rank, and by the vicious personal attacks employed when one of their members has the courage to say 'this is not right'." The release went on to say that the mission of El Grupo would be "to work with community people and environmentalists to engage in meaningful dialogue and on-the-ground projects that develop rural livelihoods and provide good stewardship of the land. Members of the group will also speak out and condemn racist and elitist actions of groups such as Forest Guardians, Southwest Center for Biological Diversity, Audubon Society, The National Forest Protection Alliance, and the Wilderness Society, whose policies negatively impact the integrity of rural and urban communities."

On September 21 we held a press conference at the state capital building in Santa Fe. Moises Morales, Rio Arriba County Commissioner, and other local activists came to support us. I read a statement detailing the controversy within the Club and our position; Ike read his own statement: "I am here today because major environmental organizations like the Sierra Club and the Sea Shepherds refuse to recognize the difference between a multi-national organization such as Duke City Lumber Company and a small local operation like La Companía Ocho, or the major whaling companies and the Macah tribe in Washington, who only want to kill several whales per year in order to maintain their culture, customs, and traditions. The bigotry, elitism, and racism exhibited by these groups under the guise of environmental protection must be exposed so that the public can decide intelligently what the true state of the forest is, as opposed to the dishonest rhetoric spouted by the leadership of these organizations." We then took out our Sierra Club membership cards and tore them up (we'd debated whether to burn them, but this wasn't 1970 any more).[75]

El Grupo press conference at the Capitol in Santa Fe. Photograph by Eric Shultz.

The no-logging policy endorsed by the Sierra Club also produced internal conflict and resignations in other major environmental organizations like Earth Island Institute. Chellis Glendinning, who at the time was on the advisory board of Earth Island Institute, was surprised when she saw that the Institute had signed onto the ad and immediately lodged a complaint with the organization. "I wrote to Earth Island and asked how it could take a public position—zero cut—that is insensitive to the basic questions of environmental justice," Glendinning said. "How could this organization, whose board of directors president is Carl Anthony, the African-American director of Urban Habitat, and whose board also includes people like Vandana Shiva, who has always worked for environmental justice, defend two conflicting policies?"[76] Others within the organization also expressed their disagreement with this public position, and an internal conflict full of racial overtones erupted.

Chad Hanson, Co-Director (along win David Orr) of the John Muir Project, a project of Earth Island Institute, wrote a letter to Earth Island's board of directors, dated October 10, 1997, in which he stated that northern New Mexico was being deforested under the guise of environmental justice and that Carl Anthony, who traveled to New Mexico to meet with *norteños*, had been taken in by this "ruse" and had made racial attacks against Hanson and other people at Earth Island Institute. He also falsely accused Ike DeVargas and other members of La Compañía Ocho of physically assaulting environmental activists. Hanson then went on at great length to assure the board that he was not a racist: "I don't have a racist bone in my body.... Yet I have been the victim of racial prejudice by the President of Earth Island Institute...because I happen to have been born with a distinct lack of pigment in my skin."[77] Hanson, along with Emily Miggins of Rethink Paper, another Earth Island Project, also wrote a letter claiming that Glendinning had a financial interest in La Compañía Ocho and accused her of being a Wise Use member. These claims were patently untrue, of course, but they put Glendinning and Anthony, both involved in social activism since the civil rights movement, in positions of having to defend themselves.

In an attempt to address this conflict within Earth Island, a group of board members and advisors formed a committee to address the issue of environmental justice. In a mission statement presented to the board of directors they wrote: "Environmental justice is a philosophy and a practice that acknowledges both the ecological destruction wreaked upon the planet by mass technological societies and also the horrific social injustices—including racism, sexism, and economic inequities—that stem from the very same systems. In today's corporate global economy, the vanguard of environmentalism becomes not just the conservation of pure wilderness against the thrust of human civilization: it becomes the fostering of human survival through non-racist, non-sexist, economically equitable community living in direct and sustainable relationship to the Earth."

Urban Habitat hosted a group of northern New Mexico community representatives and organized a demonstration at the November 6th hearings before the Ninth Circuit Court of Appeals in San Francisco, where Forest Guardians argued to further enjoin logging and grazing in Region Three. In a statement by Urban Habitat called "Fight the Corporate Destruction of Public Lands, Not Land-Based Communities," the group asked, "Why should we care about this struggle in New Mexico? Because decisions are being made here in the Bay Area without including the very people most affected."[78] In January of 1998, a group of advisory and board members, including Anthony, as well as several projects, resigned from the Institute.

Chellis Glendinning

Chellis and Ike are good friends. They'd dance together when we went to the Chamisa Lounge in Española to hear Darren Córdova or Los Blue Ventures. Ike threw a party for her 50th birthday at his trailer in Servilleta Plaza (where he cooked outside because he had no stove and washed dishes in the river because he had no running water). She gave a fundraising party for him when he ran for Rio Arriba County sheriff on a "no more jails" platform (his campaign chest totaled enough for ten signs and one newspaper ad the day before the election). When the accusations came down that Chellis had a financial interest in La Compania and that Ike had made so much money from logging that he owned two houses, we all had a good laugh. But these kinds of rumors can have grave consequences in worlds where people lack information and relationship: Chellis subsequently resigned from Earth Island Institute's board of advisors when the organization failed to respond to Urban Habitat's protest, and she eventually broke relations with other people in the national environmental community who saw her support of the Vallecitos loggers as a betrayal of her commitment to that community.

Another environmental group found itself embroiled in these issues as well. Charles Wilkinson, a well-known author and University of Colorado law professor, who was on the Board of Directors of the Wilderness Society, signed on as co-counsel for La Compania at court hearings in San Francisco where Forest Guardians was arguing that La Manga timber sale had to adhere to more restrictive forest plans and should be enjoined (La Compania was allowed to appear as a "Friend of the Court"). Explaining his action, Wilkinson was quoted as saying, "I wanted to give the Hispanics some support, and damn it, they deserve it. We in the environmental community are the ones who need to address this."[79] When contacted by the Associated Press, the national office of the Wilderness Society stated that there were no plans to censure or remove Wilkinson for his actions.

Chellis Glendinning, Ike DeVargas, and Eric Shultz.

Controversy surrounding "Zero Cut" initiatives was being played out in an international area as well. In a July 27 letter to *The New York Times*, Ralph Schmidt, Director of Forest Programs at the United Nations, wrote to support the Brazilian government's decision to open the Amazon rain forest to selective timbering, pointing out that efforts by environmental groups to completely shut-down logging oversimplifies the issue: "Harvesting forests for economic and social benefit does not necessarily equate to destroying them." He lauded the government for acknowledging that forest management was the only sustainable and economically productive option for most of the Amazon, and that it could support millions of poor people who survive by using forest resources. In a voice that could have been speaking to the situation in New Mexico he said: "These efforts [of the government] are more promising for forest conservation than a simplistic protection policy which is futile in areas of overwhelming poverty."[80]

While it was doubtful Congress would ever pass the "Zero Cut" bill, or its more moderate cousin The Act to Save Americas' Forests, which allowed some commercial cutting on public lands, the controversies these initiatives generated both within the organizations and the environmental community at large snowballed. The initiative to remove all cattle from public lands (Cattle Free by '93 just changed the date, not the intent) was put forward in various lawsuits and injunctions, particularly in the Southwest, and once again the population issue reared its ugly head, this time in the guise of a Sierra Club ballot initiative in April of 1998 to restrict immigration.

The Sierra Club: Measure A or Measure B

"Whiteness is by and large a social construct, not a description of skin color.... To be white is to fit into the social construct as the beneficiaries of European imperialism, whose relationship to the world has been one of conquest."

—Michael Lerner

"There's trouble folks, but it's not in River City and it's not about gambling: this time it's in the Sierra Club and it's about immigration. This huge organization, that most people have always perceived as representing mainstream environmentalism, is experiencing the internecine battles that only such an emotionally-charged issue can elicit."

That's how I began my article in *La Jicarita* when Sierra Club members voted in April of 1998 on a ballot that included Measure A, which read: "Shall the Sierra Club reverse its decision adopted February 24, 1996, to take no position on immigration levels or on policies governing immigration into the United States and adopt a comprehensive population policy for the United States that continues to advocate an end to U.S. population growth at the earliest possible time through reduction in natural increase (birth minus deaths,) but now also through reduction in net immigration (immigration minus emigration)."

According to Carl Pope, executive director of the Sierra Club in San Francisco, a small faction of club members forced this vote through a petition initiative.[81] In reaction to this, the board of directors, with Anne Ehrlich and Dave Foreman absent, unanimously placed on the ballot a second measure, B, which read: "The Sierra Club reaffirms its commitment to addressing the root causes of global populations problems and offers the following comprehensive approach: 1) The Sierra Club affirms the decision of the Board of Directors to take no position on U.S. immigration levels and policies; 2) The Sierra Club will build upon its effective efforts to champion the right of all families to maternal and reproductive health care, and the empowerment and equity of women; 3) The Sierra Club will continue to address the root causes of migration by encouraging sustainability, economic security, human rights and environmentally responsible consumption."

Before the vote took place, Pope wrote an editorial calling for the membership to remember the environmental mantra "think globally, act locally." By endorsing restricted immigration he warned that they would "turn the environmental mantra on its head. Instead of ameliorating the enormous challenges facing the planet" their approach would focus on symptoms rather than cause, which is global consumption and industrialization. "Erecting fences," he wrote, " to keep people out of this country does nothing to fix the planet's predicament—it's the equivalent of rearranging the deck chairs on the Titanic."[82]

Michael Dorsey, a member of the board of directors, wrote an e-mail to the board where he pointed out the fallacy in thinking that reducing immigration will address environmental degradation: "Immigration is not the cause of sprawl. Immigration is not the cause of corporate pollution. Immigration is not the cause of phosphorous loading on Eastern shore farms." While he found the Club's interest in promoting global family planning programs to address world population growth important, he pointed to our consumptive habits as the real culprit. The fact is that the U.S., which comprises 5 percent of the world's population, consumes 32 percent of the world's petroleum and plastics, produces 25 percent of the world's greenhouse gasses, and produces more solid waste than China and India combined. "It is patently absurd that here in the U.S. we can consume and destroy our resources like no society on earth and then actually use that as a rationalization against immigration."

It is more than absurd, he went on: it "smacks as elitist, imperious, jingoist, and paternalistic. But worst of all, a particularly unfeeling strain of racism lurks just under the surface."[83] And the focus of this racism is here in the Southwest, directed at Mexican and Central and South American immigrants, not the European white who immigrated at a much higher percentage rate in the early part of this century. Restricted immigration joins the list of forced assimilation policies that target affirmative action, bilingual education and ethnic studies, and have turned the U.S.-Mexico border into a war zone.

Equally disturbing was the fact that anti-immigration forces outside the Sierra Club might have been responsible for actually getting Measure A on the ballot. According to Hunter Cutting, another Sierra Club member, this initiative was funded by conservative foundations and the same anti-immigration groups in California that supported a statewide anti-immigration initiative.[84] Some of these groups made openly racist statements and acknowledged their desire to alienate the peace and social justice people from the environmental movement.

George Sessions, a well known California environmentalist and member of the Sierra Club, was one of those supporting restricted immigration. Along with several other prominent Californians, including Gary Snyder, he argued that there is a carrying capacity on the earth and we have already far exceeded it. It's easy to see why this thinking often comes out of California. As Sessions wrote in a long letter to the Club's magazine, *Sierra*, (reprinted in the California publication *Wild Duck Review*) California's carrying capacity was reached in 1965 when the population numbered 10 million (a figure he claimed "most professional biologists would no doubt agree...is close to maximum"); today it is three times that amount. He went on to say that there must be limits not only to population but limits to technology and limits to appetite and greed as well. In the same article, however, he criticized Marc Reisner's (author of *Cadillac Desert*) proposal to financially subsidize farmers to keep them from selling their farmland to development.[85] This thinking seems contradictory: Isn't saving farmland from development a limit to both technology and appetite and greed (tourism and entertainment now surpass agriculture as the state's largest industries)? Isn't a questioning of who is doing the developing and who is guilty of appetite and greed also in order? Will limiting immigration from our southern third world neighbors address either of these problems?

Dorsey pointed out that the issue had become not a question of "carrying capacity" or "quality of life" but a question of "control of which population and by whom."[86] This issue of control was really what was at stake within the Sierra Club itself as its members continued their internal battles over not only immigration but also the no-logging and no-grazing agendas. Those who saw themselves strictly as environmentalists had little use for those who were wrestling with issues of environmental justice. Dave Foreman, one of the founders of Earth First!, who served on the board of directors of the Sierra Club during this immigration controversy (billing himself as the "kinder and gentler Dave Foreman"), wrote an article in *Wild Duck Review* where he coined the phrase Progressive Cornucopianism. Called Politically-correct Cornucopianism by the unkind Foreman, he used it to describe the philosophy of "the left and New Liberalis[ts]" who don't believe that restricting immigration addresses environmental issues. He dismissed their concerns because they are essentially "socialists" who are "full of white-middle class guilt" and have no concern for wilderness and endangered species. He also attributed their motives to "Fear of Nature" and "Immaturity": "Humans are special, they all agree." He ended his article by adding that the "race-baiting hooligans of the Left...many of whom are opportunists trying to build a political base by peddling fear of racist oppression among immigrant communities" have made it clear that it's impossible to have an "honorable, decent, and fair" discussion.[87] Considering the nature of his organization, and the language he uses, it's not surprising that the Sierra Club dialogue was not about whether restricting immigration would help solve our environmental troubles: rather, it was about, as Carl Pope deplored, building fences and borders between people of color and white middle-class conservationists like Dave Foreman.

Struggle for control of the local Santa Fe Group of the Sierra Club, the state chapter, and even the national group continued after Measure A was defeated by a majority of 60 percent of the voters (15 percent of club membership). After the Santa Fe Group refused to endorse the no-logging initiative because of its insensitivity to New Mexico's land based community, Forest Guardians led an attempt to take over the group by renewing their Sierra Club memberships, disrupting meetings, and running for committee positions to determine policy. These proponents of a "take no prisoners" conservation policy, as Courtney White described it, failed to get elected to committee positions within the cohesive Santa Fe Group, but several Forest Guardians, including Charlotte Talberth, wife of executive director John Talberth (Charlotte is past president of the Forest Guardians' board and director of the Levinson Foundation), Bryan Bird, and Karen Smith, ran for positions on the Rio Grande Chapter Executive Committee and were semi-successful (Bird got elected). Talberth also tried—unsuccessfully—to become the Forest Issues Chair of the Chapter, and Smith got herself appointed editor of the Chapter's newsletter. She was particularly outspoken in her denunciation of *La Jicarita News* and *norteño* activists, and once wrote us an e-mail where she claimed Ike DeVargas owned two homes with all the money he'd made from La Companía.[88] Concurrent with these activities, their no-logging *vecinos* like David Orr ran for board positions of the national Sierra Club.

Courtney White wrote an article for *La Jicarita* about the decision of the Executive Committee of the Rio Grande Chapter to prohibit Sierra Club members from officially representing the Club at a forest issues roundtable sponsored by New Mexico Senator Jeff Bingaman in 1998. The no-logging folks on the committee claimed that they wouldn't participate in a discussion of thinning and burning to restore forest health because that's just a euphemism for clearcutting. Courtney White had this to say: "The environmental movement will only succeed if it expands its base of support instead of contracting it, as Forest Guardians, and now the Rio Grande Chapter of the Sierra Club, would have us do. The movement must embrace ideas that HELP people, especially rural people, reach commonly shared goals."[89]

Courtney White.
Photograph by Eric Shultz.

Sin Agua No Hay Vida

"I like to use the word maintain rather than preserve when I'm talking about *acequias*. The word preserve has the connotation that something is at an end."

—Estevan Arellano

As a result of the work Mark and I had done regarding the proposed Sipapu Ski Area expansion, and the efforts we made to organize a federation of *acequias* in our watershed, the New Mexico Acequia Association asked me to serve on its board of directors in 1998. Water, of course, is the issue that underlies everything—logging, grazing, sustainability, economic development, cultural tradition: *Sin agua no hay vida*, without water there is no life.

The *acequias* of northern New Mexico are an engineering feat, especially when you consider they were constructed with hand tools and without any kind of engineering equipment to measure land grades and flow rates. All the indigenous cultures of the area used *acequias* for agriculture, from the Puebloans to the Spanish colonists to what became the Indo-Hispanos, who in the seventeen and eighteen hundreds expanded and improved the *acequias* as the lifeblood of the north. Not only do they carry water, by gravity flow, from the valley rivers to an extensive system of fields and orchards in just about every village of northern New Mexico, they also define the social and political nature of the community. Very few of the villages they serve have any kind of governing body other than the *acequia* commission, which administers the ditches. From the early days of subsistence farming to today, when second and third generation family members still use the water to irrigate their fields of alfalfa, vegetable gardens, and orchards, the *acequias* are what keep people tied to the land and to each other. They also serve the land in an ecological way by recharging the groundwater aquifer and supporting wildlife habitat with their extensive riparian corridors.

Water law in New Mexico, as all over the west, is complicated. Specific law that deals with *acequias* is even more complicated, but the fundamental issue in *acequia* water management derives from fundamental conflict: should *acequias* be managed as they have since their inception, based on Spanish and Mexican law, as a community resource that benefits all *parciantes,* or should they be managed as a private property right, under European law, that can be severed from the land and bought and sold in a market economy. It is a conflict that penetrates to the very core of every battle being waged in northern New Mexico, as local communities struggle for the autonomy to sustain their culture and economies in the face of overwhelming federal and global control.

While 80 per cent of the state's current water use is agricultural (and much of that is actually consumptive use by the riparian corridor and 50 percent is return flow), what these forces see as the "highest and best" use—meaning that which has the highest economic value—of New Mexico's water resources is urban, industrial, and recreational. The intrinsic values of rural communities that the *acequia* system supports are being commodified for their economic significance. A particularly alarming example of this trend is the *Waste Management Study: Upper Rio Grande River Basin*, commissioned by the Western Water Policy Review Act of 1992, which was mandated to define the federal role in water policy for the west. The report, among other things, states that "We recommend federal agencies in the Basin do more to mitigate the constraints to competition that keep water and other resources in low-value uses while high value demands go unmet." It goes on to say that *acequia* associations and irrigation and conservancy districts have exercised undue influence on legislation pertaining to water distribution in the state and that the Rio Grande Compact governing distribution on the Rio Grande between Texas, New Mexico, and Colorado "...reflects the agrarian economy...that existed at the end of the 1920s, not today's highly urbanized economy." The report obviously uses only market values, not human values, and in essence frames economic value from a capitalist—and colonialist—perspective. Even from an economic standpoint the report ignores the fact that the main source of income here is tourism, and tourism is generated by the rural/agricultural nature of the state. There is no mention of the potential for value-added agricultural such as niche crops, grass-fed beef, or wheat, which was formally grown in northern New Mexico and referred to as the breadbasket of the state. The report also implies that *acequia* associations have the same political clout as organizations such as the Middle Rio Grande Conservancy District, which governs water use in the Albuquerque area, when in fact *acequia* associations are underfinanced and largely disenfranchised.

It's not only government agencies going after *acequia* water, of course. Just as they have used federal laws such as the Endangered Species Act to shut down logging and grazing on public lands, environmental groups have threatened to use the same laws to override state and local water authorities. Because water in New Mexico is already over-allocated, meaning the only way to acquire water rights to maintain minimum stream flows is to buy and change their purpose and use, *acequia* water rights are seen as the most vulnerable. Groups like Forest Guardians, who have threatened legal action to break the Rio Grande Compact, fail to understand that they are helping create policies that commodify water in the free market. The transfer of *acequia* water to instream flow for recreational purposes such as rafting, or for environmental protection of endangered species, is in direct opposition to keeping water within communities and tied to the land. As Antonio Medina, president of the New Mexico Acequia Association, put it at the annual meeting, "No one owns the water. It is a sacred resource. But as *parciantes* we will fight for our right to *use* the water as we have always done, to sustain our way of life."

The method we have to fight water transfers is guaranteed under state law: if you can prove, before the Office of the State Engineer, that a transfer and/or change of use would impair existing users, is contrary to conservation of water within the state, or detrimental to the public welfare, the State Engineer can deny the transfer. This "public welfare" part of the state statute was not added to the law until 1985 and has yet to be tested in court. But it is what many hope will prove to be the definitive way *acequias* will be able to protect their water rights. It is what we hoped to argue in the Sipapu Ski Area proposed water transfer, before it was discovered that the ski area had no water rights to transfer. José Rivera, a native of northern New Mexico who is a professor of public administration at the University of New Mexico, wrote a paper for the Rio Pueblo/Rio Embudo Watershed Protection Coalition arguing that ski area expansions bring "inappropriate" development to rural communities and are therefore detrimental to their public welfare. The New

Mexico Acequia Association has taken the position that we must oppose any water transfer application that proposes to change use from agricultural to commercial.

We obviously have an uphill battle before us. But there have been, and will continue to be, wonderful, dedicated people committed to the task: members of the New Mexico Acequia Association like Josie Lujan and Paul Garcia, who dedicated their entire adult lives to the protection of the *acequias* that reared them; David Benavides, a legal aid lawyer and William Gonzales, a para-legal, who specialize in water law and represent *acequia parciantes* in their water transfer fights; Doug Wolf, a public interest attorney who took on Intel Corporation when it attempted to transfer agricultural water rights to its plant and won; Peter White, a former State Engineer attorney who now fights transfers for *paricantes* free of charge; and my neighbor, who lives in Santa Fe but comes to the village, after work, when it's his turn to water the fields at his grandmother's house, where he was raised. The strong attachment to the land exemplified by my *vecino* remains intact; to keep the water on that land is our duty and our right.

"Zero Cows"

> "Seldom does anyone outside of the ranching community even recognize grazing as a legitimate use of public lands. But few have been as vocal as the so-called environmentalists have been in trying to force the outright elimination of grazing from public lands. I say so-called environmentalists because in my opinion they are often so intent on eliminating grazing that they have failed to consider the long term, big picture consequences of their anti-grazing blinders."
> —George Maestas, northern New Mexico rancher

Remember the slogan, "Cattle Free by '93?" I remember seeing a magazine put out by Earth First! with all the terrible pictures of degraded rangelands overrun with "welfare cowboys" cattle. I don't think any of the pictures were taken in northern New Mexico, or if Earth First!'s notion of a welfare cowboy included George Maestas, lifelong resident of Rodarte, heir to the Santa Barbara Land Grant, and national forest permittee who grazes fewer than 20 cows on lands that once belonged to his ancestors. I'd say their use of the term welfare reveals more about them than it does about ranchers: environmentalists promoting the agenda of no grazing on public lands join the ranks of reactionary Republicans who take the position that any kind of government funding—to help the economically oppressed or support rural community livelihoods—is deplorable. Considering Earth First! founder Dave Foreman's history as a Young Republican for Nixon, I'm probably not so far off the mark.

In a report called *Of Cows, Culture, and Continuity: An Environmental Justice Argument in Support of Public Lands Ranching in Northern New Mexico,* author Ernie Atencio provided faces and names of the people of northern New Mexico who would suffer from absolutist policies like "Zero Cow." The Santa Fe Group of the Sierra Club funded the writing of this report, a courageous challenge to the national Sierra Club's referendum that called for the termination of public lands grazing (the referendum was soundly defeated). Atencio also supplied statistics to support his argument that it is not the small-scale ranchers of northern New Mexico who benefit from public subsidies or are responsible for the land degradation that does result from over grazing on public lands. For instance, nationwide, the top 10 percent of BLM grazing permit holders control 65 percent of all livestock on BLM lands. None of that top 10 percent is in New Mexico. Compared to the 100 permittees in Rio Arriba County who share a quarter of a million acres on 11 adjoining allotments on the Santa Fe National Forest, 16 Elko, Nevada ranches are larger than 100,000 acres.

But beyond the facts and figures that support an argument for public lands grazing in northern New Mexico is the reality of *norteños' querencia*, the affection and longing they have for the land: "The more you work the land, the more you get to love it, because your heart is in your land." This is a *dicho* from Andie Sanchez's grandfather that expresses the connection these third and fourth generation ranchers feel towards these lands that continue to sustain them, despite the loss of the land grants. They maintain a proprietary attitude towards the land that engenders a land ethic of responsibility and stewardship. Atencio cautioned readers not to "romanticize land-based Hispano culture as a paragon of environmental harmony and sustainable resource stewardship" because no culture on earth "can claim a history of perfect, benevolent stewardship. Nonetheless, an ethos of restraint is and has been the general guiding principle of resource use, or cultural ecology, in northern New Mexico for centuries."[90]

It is this attachment and restraint that have created the localized agricultural enterprises in northern New Mexico that are not only a link to the past but can serve as models for sustainable communities in the future. Atencio listed some of the agriculturally based enterprises that provide this link: Ganados del Valle, Tierra Wools, Pastores Collections, Sangre de Cristo Growers Cooperative, Las Humanas Cooperative, Madera Forest Products, and La Montaña de Truchas Woodlot. He listed the innovations in range management that are revitalizing ranching in northern New Mexico: holistic range management, that Virgil Trujillo, rangeland manager at Ghost Ranch utilized; the herding employed by rancher Joe Torres on the Valle Vidal, and the Valle Grande Grassbank. These are all examples of work that has been achieved through collaboration and consensus, concepts that people who work on the ground know are essential to maintaining healthy economies, communities, and resources. Atencio ended his report with this statement: "If we care about genuine environmental justice and about setting right historic injustices, then supporting local Hispano self-determination through appropriate and sustainable economic development is a move in the right direction. Finding a way to insure traditional-use access to ancestral lands and resources, though sure to stir more controversy among the old guard of the environmental movement, moves us farther in the right direction."

Unfortunately, the "old guard" environmental movement continues to do everything in its power to move us in the wrong direction. While the cows are still out there on forest allotments, their numbers are diminishing, due to lawsuits filed by Forest Guardians (now known as WildEarth Guardians) and the Center for Biological Diversity and new, more restrictive Forest Service guidelines. John Horning, then water specialist with Forest Guardians, told me point blank that "I have no illusions about fighting the corporate world. I'm not. I just want to see streams and watershed in the Southwest in better shape."[91] If it were that simple, there would be no controversy: we all want streams and watersheds in better shape. Unfortunately, when folks like George Maestas are included in lawsuits to remove cattle from public lands, it is the developers, high-tech recreationists, and corporate interests that will step in to finish off any vestiges of biodiversity and open space.

Those who differentiate between global, corporate interest and small, traditional ranches that do business on public lands have been successful in establishing dialogue among open-minded environmental groups, government agencies, and most importantly, the people who work the land. In June of 1998, the Quivira Coalition and the Rio Pueblo/Rio Embudo Watershed Protection Coalition organized a one-day workshop to discuss the unique cultural and ecological needs of *el norte*. Held in the Peñasco High School multipurpose building, it was quite a sight to see: Hispano ranchers; mainstream environmentalists; Santa Fe foundation directors; Green Party members; bureaucratic agencies including the Forest Service, the Bureau of Land Management, and Taos Soil and Water Conservation District; a land trust representative; and some land restoration folks from Colorado listened to speakers, saw a slide show about what fire suppression has done to grasslands, ate pizza, and all went out to a thinning and prescribed burn area to see just what's going on with forest and range restoration efforts.

Virgil Trujillo, then rangelands manager at Ghost Ranch, a

Presbyterian-held guest ranch and conservation organization, moderated the Peñasco workshop. As a land grant heir whose father previously grazed his cattle on what is now Ghost Ranch property, Virgil established an immediate rapport with the Peñasco area permittees and spoke of the holistic management style he employed at Ghost Ranch, which provided winter rangeland to area permittees. Andie Sanchez and George Maestas of the Santa Barbara Grazing Association discussed some of the problems they faced as small-time cattleman on the forest allotment: loss of rangeland due to forest encroachment; lawsuits that imperil grazing on public lands; and new regulations imposed by the Forest Service. Maestas read an impassioned statement that tried to put the dilemma into perspective: "It seems that every week cattle and ranchers are blamed for some new environmental catastrophe. It's gotten to the point that I wouldn't be surprised to hear us blamed for problems with the Japanese economy or failure of the peace process in the Middle East." Another rancher, Palemon Martinez from Taos, who is a member of the Northern New Mexico Stockman's Association, spoke about the ways his group was trying to address degraded rangelands by supporting the acquisition of grass banks, public or private lands designated as temporary grazing allotments, to provide opportunities for northern New Mexico ranchers to rest and restore their forest allotments. His group had been supportive of The Conservation Fund's Rowe Mesa grass bank and was a member of a coalition pursuing public acquisition of the Jemez Mountains Valle Grande, a huge tract of forested land that was being sold by the Texas ranching family that has owned it for many years. Martinez pointed out that forcing permittees off public lands would break these people's traditional ties to the land and ultimately result in more development and the destruction of our rural/agricultural communities. He asked the group, "Would you rather see 50 cows or 50 houses?"[92]

A scientist was on hand to provide an understanding of how natural ecological processes have been subverted by bad management practices and have left us with overly thick forests and degraded watersheds. Craig Allen, who works for the United States Geological Survey at Bandelier National Monument in the Jemez Mountains, has studied the history of fire over hundreds of years and has charted landscape changes over the last 55 years. Because of public lands agencies' long history of fire suppression, the montane grasslands, which have existed for thousands of years, have steadily been disappearing at the rate of about one per cent a year over the last 50 years. Allen's recommendation for appropriate restoration projects include cutting and burning of trees out of invaded grasslands and meadows, thinning and prescribed burning on ponderosa pine forests to reduce the density of understory trees, and thinning younger piñon and juniper from thick woodlands. These efforts "could help resolve persistent range management conflicts on public lands by providing additional grazing capacity on upland settings away from the environmental conflicts associated with grazing in riparian areas."

Author and environmentalist Bill deBuys, who worked for The Conservation Fund, which operated the grass bank on Rowe Mesa, was the last speaker. He told everyone how his area could provide year-round grazing for up to 325 cattle from other national forest allotments that need to be upgraded. Permittees who took advantage of the program could place their cows on the ranch allotment for short or extended periods while they made efforts to upgrade their own allotments.

After lunch, Ben Kuykendall, wildlife biologist on the Camino Real Ranger District, led the group on a tour of the Borrego area outside Peñasco (different from Borrego Mesa on the Santa Fe National Forest) where the Forest Service had thinned and burned. Everyone could see that the forest canopy had been reduced, allowing for the growth of native grasses and encouraging the growth of larger trees. The area was first opened up for community fuelwood gathering and then burned at two different times to regenerate growth. It was good for everyone to get out in the woods, take a look at the existing conditions, and see how the projects were actually changing those conditions. As a result of the workshop, organizers got together with the interested parties and began working collaboratively to improve forest health and support the Santa Barbara permittees in their goal to upgrade their allotment. Andie Sanchez and George Maestas, along with almost all of their fellow permittees, committed to working with whomever was supportive of the project. The New Mexico State Environment Department, which was administering an Environmental Protection Agency grant to remediate non-point source pollution in our watershed, committed more than half of its funding to thinning, burning, fencing, and trail rehabilitation in the allotment. The Conservation Fund agreed to let the Santa Barbara permittees use the Rowe Mesa grass bank for at least two years while the allotment was rested and restored. The Quivira Coalition committed funds to monitor the restoration projects from beginning to end, and the Rio Pueblo/Rio Embudo Watershed Protection Coalition applied for funding to rehabilitate and construct upland water resources for both wildlife and cattle to prevent riparian area degradation. It was the kind of project that made everyone—community members, permittees, environmentalists, bureaucrats, and foundations—very happy; except those environmentalists whose agenda is "Zero Cows."

Tomás

"I would propose that we consider the human population as an indicator species of ecosystem health."

—Marco Lowenstein

During most of my years in El Valle Tomás was the unofficial mayor (he died in 2009). While he officially functioned as one of the *acequia mayordomos* and took his turn as the *mayordomo* of the Catholic church, there really was no official mayor. But he was it, no question about it. He had the biggest woodpile, owned three tractors, one of the few balers, was a cousin of our county commissioner, and knew everybody's business, good or bad. Fortunately, he was also my next-door neighbor, and with that came the privilege of a special intimacy and benevolence that opened the door for our acceptance by the rest of the community.

I remember the first time I met Tomás. I already knew who he was, described by the man we bought our house from as "the nicest Republican chauvinist you'll ever want to meet." (The reason Tomás was a Republican was that the Democratic precinct chair was already occupied when he decided he wanted a seat.) He came down the driveway one day, to let us know we could have the water that afternoon and immediately asked where Mark was. I introduced myself in Spanish, and this huge man in a cowboy hat and work boots gently shook my hand but never really looked me in the eye. He reluctantly relayed his information to me and quickly departed. The ditches were obviously not only a Hispano domain but a male one as well, and I could sense his unease communicating with a newcomer who was both female and white.

Over the years the nature of my relationship with Tomás became so warm and relaxed that the initial memory I have seems almost of a different person. There was nothing Tomás wouldn't do for us: he babysat our son Max, taking him for rides in his pickup to check his cows; if our car broke down, he came and got us; if we needed any ditches or fence post holes dug, he brought over one of his machines to do the job; if we ran out of hay, he provided it. Conversely, we took special care to return the favors: we provided him with fresh eggs; we helped him cut and bale hay; we baked him cherry pies and shared the juice we made from our apples; we read any complicated letters from lawyers and bureaucrats that he might have trouble deciphering. In fact, he considered our garden and orchard his as well. Once he'd provided the manure in the spring, he'd keep an eye on how things were going and then come over on his four-wheeler to pick the plums hanging on the driveway trees, work his way down to the sweet cherries in the orchard, and then into the garden for a few red, juicy tomatoes.

There was an unspoken law between us that any favor asked would be granted. It was based on an understanding that the favor would not be unreasonable, that it was necessary and not frivolous. Sometimes, because of our cultural differences, there might be a certain shaking of the head, a muttered, "those crazy *gringos*" or "that *loco*," but we acceded to the other's wishes, and we wrote it off as what you do for a friend, pure and simple. I loved Tomás unreservedly, despite all the judgments I brought to bear. I hope he loved me the same way.

Tomás was born in the house next door to us, which became a ruin full of mice and snakes and whatever else wanted in out of the rain. Like many older northern New Mexico houses, it was comprised of a series of rooms connected to each other railroad-car style, with an outside entrance to each room. When he was a young boy, the family moved into his uncle's house just up the road, the same style, but a little larger after a pitched roof was added, providing a sleeping loft upstairs. There his parents raised seven children, two girls and five boys, three of whom continued to live in the village. One brother ran the local store, while the other one moved back after retiring from a factory job in California. Like many men of his generation, Tomás worked off and on for years as a sheepherder in Wyoming, a miner in Colorado, and in a canning factory in California. Consequently, their wives took care of the children, cows, fields, and gardens while they were away. Tomás and his wife, Ermenita, originally from Chimayó, had two children, Fred and Helen, who both live in the village. Unfortunately, unlike most men and women of their generation, Tomás and Ermenita were divorced after twenty-nine years of marriage, and after moving back to the village where she was raised, she died of cancer in the 1980s.

Tomás lived the bachelor life with his son Fred, in the frame-stucco house he eventually built next door to the two family homes. Tomás did the cooking, Fred did the cleaning, and they both entertained women friends whenever they wanted. Tomás obviously enjoyed his bachelor life and was never without a special woman friend, who he took dancing, to funerals and weddings (Tomás attended a lot of funerals, guaranteeing that his own would top all in attendance), and to all the village functions. Tomás loved to dance. Mark and I would go with him and his girlfriend to the Sunday dances at Sider's in Las Vegas, where he and I danced together to the fast songs so he could swing me around the floor as hard as possible. A very important criterion for choosing a woman friend was that she must like our village and enjoy spending time here. I once asked Tomás what happened to the woman from Santa Cruz who he had been seeing for quite a while. His response, "She always wanted me to come down there to go out to eat or go dancing or see her family," made it clear that his responsibility to his village superseded all other attachments.

Orlando and Tomás Montoya.

If we do as Marco Lowenstein proposes, Tomás would be the human indicator species of northern New Mexico. He was defined by what he did, by who he was in his community, by his sense of what a community is. Some may pass judgment upon him as essentially a benevolent dictator. He bumped heads with a few younger men who, returning to the village after many years away, challenged his authority. Some may pass judgment upon him by criticizing his reliance upon cows for his livelihood. I certainly questioned his monetary rate of return based on the amount of work he put in (as a writer, however, I'm probably in the same league). But like most other ranchers up here, he ran fewer than thirty cows, had plenty of land to graze them, and owned a national forest grazing permit cooperatively with several neighboring ranchers. And raising cattle was really only part of his livelihood. He drove a school bus for many years, was a bus contractor for the Forest Service, and cut everybody's hay.

His generation may be the last, at least in our village, that will be involved in the cattle business. Fred may continue to run a few cows, but not to the extent Tomás did. His daughter Helen drives a school bus, his nieces and nephews have moved away to work in Española or Los Alamos, and everyone else up here has some other form of employment to sustain them. But really, it was Tomás's example of community spirit and sense of place that is the indicator of our collective health. Those of us who live here, whose families have lived here for centuries or who have moved here because they *wish* their families had lived here for centuries, must be able to make a living that defines us as caretakers of the land, the community, and the soul of this wonderful place. And what that requires is working with the land, knowing it intimately, knowing how we relate to it, knowing what is necessary to sustain it. In our small corner of the world, as in many rural areas around the world, I think it is important to hold on to the jobs that do just that: growing something, be it more vegetables than hay; feeding something, be it a llama or a cow; cutting *vigas* and *latillas* as well as sawtimber; heating with passive solar heat as a back-up to firewood rather than in its place.

These jobs ensure a connectedness to each other as well. I can sit in front of my computer all day, churning out words and thoughts and ruminations on all kinds of things, many of which are nebulous or meaningless to anybody but me. But when it's time to go get a load of firewood, get some manure for my garden, or fix the fence, I have to rely on my friends and neighbors for help. And that human contact, already so severely limited by cars, computers, fax machines and modems, is, paraphrasing Thoreau, the preservation of the world.

The Roundup

"You cannot save the land apart from the people or the people apart
from the land. To save either you must save both."
—Wendell Berry

Tomás had a Forest Service permit to run about 20 to 30 cattle on two grazing allotments in the Carson National Forest. He was part of an association, as are most permittees in northern New Mexico, whose total herd is about a hundred cows. Tomás was a man of tradition, so when it was time to move the cattle from the early summer allotment to the mid-summer allotment, he liked to move them the traditional way: on horseback down the middle of the road. Only now the road is a paved highway full of tourists and the cowboys riding the horses included *gringas* like me.

I hadn't been on a horse in about ten years but for some strange reason I thought I'd be able to ride on this roundup with impunity. Tomás would ride his huge horse, Primo (Tomás, a huge man himself, required a huge horse) while I would ride his other no-name horse (that we shall call Blanca for lack of anything better), a lazy, middle-sized horse that fortunately NEVER BUCKED. We loaded the horses into Tomás' trailer and picked up Bill deBuys along the way, who was late because he'd borrowed an Arabian horse from Crockett Dumas, the district ranger, and had spent the entire morning struggling with the horse-shoer to get shoes on this fine-spirited horse that preferred to go shoeless.

We drove from the village down the paved highway about six miles to the forest road turnoff that led to the allotment. Then we drove along the dusty dirt road for I don't know how many miles (it was a lot of miles) to the corral at the base of the mesa-top allotment. We would bring the cows down to the corral to count before driving them to their new allotment, which was back the way we came, several miles beyond our village. I knew all of this, conceptually, but I somehow repressed any notion of exactly what this might mean, mileage and time wise, or I would have quit before we began.

The group assembled at the corral included the two *gringos* (Bill and me), the rest of Tomás' association (a couple of them didn't ride horses, but followed in their pickups), and their cowboy buddies, one of whom was also from our village, the soccer coach at the local high school. Eliu gave me grief about my horse all day long: "I wouldn't ride that nag if you paid me" and "I don't trust that horse for nothing." I, of course, appreciated the nag factor, and Tomás kept telling Eliu, "What do you mean, I can do anything to that horse. He's a good horse." This was from a man whose whole life was predicated on the fact that really, one could *never* completely trust a horse, and he was right. But he didn't want anyone disparaging *his* particular horse.

We mounted and immediately started climbing to the mesa top along a steep, gravelly trail that caused all the horses to slip and slide and caused me to hold on for dear life. I followed right behind Tomás, basically giving the horse the lead and praying that I wouldn't fall off before we even got started. I survived the climb, and once on the mesa I could relax a little and enjoy the scenery, which was spectacular, with views extending in all directions towards the mountains behind us and the river valley ahead. There seemed to be plenty of grass left on the mesa, and Tomás explained to me that they had to move the cows because there wasn't enough water: the trick tanks dried up during the mid-summer drought and the cows had to go higher up the mountain where water flowed all year long from natural springs.

We divided into teams spread out across the mesa top to look for the cows. I stayed with Tomás and enjoyed the gentle trot through the grass towards the far tanks where the cows were assumed to be. Everyone converged on them from their various directions, and the cows obediently headed across the mesa towards the trail to the corral. We stopped at a still-filled water tank in a clump of oak trees, where the cows and horses took a drink, then headed them down the side of the mesa where they took off in all directions as they found their own way down the steep embankment. I stayed to the rear and once again prayed that the horse could find its way down the trail with me on its back.

Back at the corral I got to rest my already weary legs while everyone milled around trying to separate everyone's cows so they could take a count. I watered the two dogs that had been brought along—Tomás' blue-heeler

Cuete and a red-heeler that was apparently along just for show, as he rode in his owner's pickup the entire day. Finally, we mounted and started moving the cows up an arroyo towards the road. I'd brought my camera along and took some pictures along the grassy arroyo. By the end of the day I'd lost my lens cap and coated the camera with a layer of road dust that still lingers in nooks and crevices.

Once we were on the road, an errant cow headed off in the wrong direction and every damn rider except Bill, Eliu, and I went off to chase it. Now I thought this was a little strange, that it took this entire group of experienced cowboys to chase one cow, but who was I to question the ways and means of a cattle drive. So there we were, two *gringos* and a soccer coach behind a hundred head of cattle going every which way through the piñon/juniper landscape. Fortunately, Bill had done this before and was an excellent rider (I would later question his sanity, wondering why in the world he would choose to do it again). Eliu had grown up on a horse, so he supposedly knew what he was doing as well. I spent the entire ride trying to keep my horse from getting *ahead* of the cattle or keep from poking my eye out as we went crashing through the trees after wayward cows. Half the time I didn't even know where Bill or Eliu were: Bill spent a lot of time trying to keep his horse on four legs, while Eliu kept riding on ahead to make sure gates were open or some such thing (I later figured he was just trying to stay ahead of the dust). This is apparently what all the rest of them were doing as well, as they failed to rejoin the group until we were about a mile from the highway. I finally realized that the cows actually knew where they were going, and if we'd just let them out of the corral, eaten a casual lunch and followed them at a leisurely pace in the pickups, we would have all eventually met at the highway.

Instead, Bill and I ate their dust and fought piñon branches for many miles down the road. Eliu reconnected at some point and finally, after several hours of riding, the rest of the cowboys suddenly appeared, chattering and laughing and casually letting the cows run any way they wanted through the trees.

"Where the hell have you been, Tomás?"

"We had to chase that dumb cow."

"It took all four of you to chase one dumb cow? Bill, Eliu, and I have been out here busting our butts trying to keep these cows on the road."

"Oh, these cows know where they're going, all you have to do is ride along and give them a holler every now and again."

"Then why the hell am I out here doing this?"

"Because now the fun starts, we get to ride down the highway for all the *turistas!*"

At least there wasn't any more dust. There were cars instead—many of them, all over the road, along the shoulder, trying to get around us, filled with people trying to take our picture, pointing and laughing in the general chaos. I just tried to do what I was told, although I didn't particularly like the cowgirl (daughter of one of the permittees who had joined us at the highway) who was doing all the telling. And to top it off, she was riding Tomás' horse while he rode in one of the pickups (Cuete was in the pickup, too, completely worn out). After a couple of miles I noticed that Tomás wasn't the only one who had taken a break. Eliu had traded with his wife Penny, who had also recently joined us, two other permittees had give over their horses to other fresh blood, and one had even tied his horse to the back of the pickup in which he was now riding. It seemed that the two *gringos* were the only ones on their original mounts.

I rode over to Tomás' pickup.

"I'm tired, I want to take a break."

"We don't have nobody else to ride your horse."

"Then why can't I just tie her up to the pickup like that other guy. You don't need me."

"Yes we do. You have to ride your horse, *mi hi'ta.*"

My legs were killing me, I had a cramp in my foot, and the sun beat down hotter than ever on my poor *gringa* face, despite my straw hat and sunscreen. I complained vociferously to Bill, but he just shook his head and said, "We're just the *peones* here, so you might as well accept it."

We finally stopped for lunch at a wide spot in the road, which we shared with the cows. The cowboys and their wives had kindly sent along chicken, rolls, salads, beer, and sodas. I sat in the only shade provided by the side of one of the pickups and again begged Tomás to let me quit.

"We don't have nobody else to ride your horse."

I almost cried when I had to remount. I just about fell off when the horses started fighting for position at the water hole. I got yelled at in Las Trampas for getting too far ahead of some of the cows. My friend David, who owns one of the roadside stores in Trampas, laughed at me. We kept on riding, one more mile to our village turnoff, then one last mile through the wide road-side arroyo to the next village turnoff that led to the allotment. As soon as the cows were safely off the highway I got off Blanca, tied her to the first pickup I found and announced to Tomás, "I quit. I'll walk home if I have to." I couldn't have walked ten feet, let alone the three miles it would take to get home.

"We all quit. This is the end of the line."

"But the pasture is about six more miles up the road."

"Yeah, but the cows know how to get there. We just turn 'em loose and they find their way home."

He handed me a beer, and then said, "Want to go dancing tonight?"

I went over by the side of the road to be miserable by myself. I could see that Bill was just about as exhausted as I was. When Tomás told him he was going to take the pickup up the road later to check on the cows, which meant that Bill and his horse would have to wait a couple of hours for a ride home, Bill decided to ride back to the village, cross-country. I never could have done it, and he told me later he wished *he* hadn't done it because by the time he got home he had such bad leg cramps he couldn't drive his car back to Santa Fe until the next morning. Then, just before tears, Tomás' brother showed up in his Bronco and took me home.

"Boy, you are one crazy *gringa*. What did you go and do that for? I'd never ride that many miles after those stupid cows."

"Yes, I agree, I am one crazy *and* stupid *gringa*, same as the cows."

Tomás called the next day to see if I could walk.

"You thought I'd be so sore I couldn't get out of bed, didn't you. Well, I'm feeling just fine, *viejo*, but I'll tell you I'm never going to ride on your roundup again."

"We'll see, *mi hi'ta*. We'll see."

La Jicarita Editorial, September 2000

When *Uncommon Ground: Rethinking the Human Place in Nature* was published in 1996, many environmentalists interpreted its message—that "nature" is a human idea, with a long and complicated cultural history—as a hostile attack on the environmental movement. The book, an anthology of essays by well-known teachers of environmental science, history, sociology, biology, etc., was born of a conference at the University of California at Irvine in 1994, where participants came together to look at environmental problems from a humanistic, interdisciplinary perspective. The book's editor, William Cronon, addresses the attack by environmentalists in the book's foreword: "The criticisms we offer—whether of environmentalism in particular or of American ideas of nature in general—are intended to encourage greater reflection about the complicated and contradictory ways in which modern human beings conceive of their place in nature.... At a time when threats to the environment have never been greater, it may be tempting to believe that people need to be mounting the barricades rather than asking abstract questions about the human place in nature. Yet without confronting such questions, it will be hard to know which barricades to mount, and harder still to persuade large numbers of people to mount them with us."

This question is what many of us in *el norte* have been asking for the last few years: has the pendulum swung too far in our reaction to what humans have "wrought" on the "natural" world so that many environmentalists, rather than helping find a middle ground to lighten this heavy hand, encourage the notion of man's separateness from nature? The failure of these environmentalists to deal with human social issues because of this idea of separateness results in policies like Zero Cut and the Wildlands Project, which define any human touch, be it corporate or community, as inherently bad.

Cronon, who is a professor of history, geography, and environmental studies at the University of Wisconsin, talks about this dualism in his chapter "The Trouble with Wilderness." By defining wilderness as something that stands apart from humanity, environmentalists often encourage conflict between those who value wilderness and those who are too preoccupied with solving their own environmental problems of toxic poisoning or loss of land, water, and land-based employment: *en otras palabras*, the poor. "This in turn tempts one to ignore crucial differences among humans and complex cultural and historical reasons why people may feel differently about the meaning of wilderness," Cronon writes.

By not addressing issues of environmental justice, environmentalists lose their moral ground. Here in northern New Mexico, both Native American and Hispano traditional communities have been dependent upon their surrounding forests for hundreds of years; when these common lands became public lands, they lost most of the "inhabited wilderness" that sustained them as societies. As rural *el norte* becomes more vulnerable to global economics it is the responsibility of all of us who value wilderness to extend our concern to the sovereignty of inhabitants of wilderness. And there are many groups doing just that: community forestry groups like La Montaña de Truchas and Madera Forest Products, and the coalition that is working to rehabilitate the Santa Barbara Grazing Allotment (including the Quivira Coalition). These are our issues of environmental justice.

We believe that there can be a more holistic way of integrating both cultural and natural landscapes where conflicts can be resolved. Too many environmentalists speak in absolutes, devising ethics in the abstract and applying them across the board. We must be able to find a middle ground based on the ethics of place, incorporating the social, ecological, political, and economic realities of each situation. In her chapter "Constructing Nature: The Legacy of Frederick Law Olmsted," Anne Whiston Spirn sees this middle ground as that between John Muir's idea of nature as "temple" and Gifford Pinchot's idea of nature as "workshop," a balance between reverence and use. Just as the loggers and environmentalists of the Pacific Northwest must come to terms with the best way to save ancient forests and the small-town economies dependent upon logging, so too must we in *el norte* commit to keeping people on the land through sustainable use of our forests and watersheds. Another contributor to the book, James Proctor, in his chapter "Whose Nature? The Contested Moral Terrain of Ancient Forest," calls these efforts "inclusive environmentalism, one uniting it with other social movements in a common moral cause: to help create a more livable world for all of us, humans and nonhumans alike."[93]

Epilogue

> "You may want to stage your own signifying practices to enrich, combat, modify or transform the effects which others' practices produce."
>
> —Terry Eagleton

It was 20 years before I went back to Placitas to see my old house. Or rather I went to see what was left of my old house. Actually, all I saw was the *place* of my old house. My neighbors had told me years ago that the folks who bought the house built onto and around it, kind of like folks who build onto or around a trailer to turn it into a more solid construction; but my house, an adobe for god's sake, was nowhere to be found.

Neither were the fruit trees, the cottonwoods, the native grass, the rock gardens, the vegetable and flower gardens Mark and I spent fifteen years cultivating, carefully meting out the precious water that is pumped from hundreds of feet below ground. I guess when they took off the second story, added on rooms (including one at the west end with a turret) and then put on a different second story, the vegetation got in the way. This one now resembles a lot of the houses built over the years I've been away: solar adobe style, boxy, oversized, characterless.

But some of our handmade houses are still there: Cathy's Dome out on the mesa; Rumaldo's *ranchito* on lower Las Huertas Creek; the crumbling adobe in the village where Mark and I lived while we built our new house; the triple dome and solar dome (where we fell in love) in Dome Valley; and Daisy's carefully crafted compact house on a windswept hill. Things change, of course (just ask the land grant community). People build houses. Developers build many houses. In El Valle we suddenly have second homes. But Placitas is emblematic of a consumer culture that is the basis of this country's power and wealth. We became the "primary engine of capital accumulation" (David Harvey), which vastly increased the infrastructure necessary to support this kind of massive suburbanization. The housing boom in the 90s exacerbated the overbuilding and created the upper middle class culture that now dominates Placitas. Now, in the 2000s, there must be a slew of foreclosures and some nail-biting developers who won't survive the housing bust, but the damage has been done: loss of the commons, validation of yuppie culture, and the mining of the already limited water resources of a semi-arid environment. (I recently ran into Lenore Goodell at the wake of an old Placitas friend. She told me she can no longer garden and can barely keep her trees alive because there's not enough water.)

My friends who stayed say they drive up the road alongside "Homesteads" and "Ranchos" with their eyes pointed straight ahead, following the route home without looking. When they take their morning walk up the hill and see another road bladed through the desert, they look away. They're attached to their homes and their history in the crazy, mixed up place Placitas was and is. I prefer the "was."

After we'd moved to El Valle the editors of a monthly community newspaper in Placitas (that deals mainly with community events, the arts scene, and some land-use issues) called to see if Mark and I would be interested in writing an article for them explaining how our "environmentalism" had changed since we moved to *el norte*. This is what we wrote: "If by our 'environmentalism' you mean our politics, we would say they haven't changed at all. We are both products of the 60s, reared on the civil rights movement, the Vietnam War and New Left, and the women's movement. As we traveled this political path we were forced to address issues of race, class, and gender; if we add environmentalism to the list of movements, we must bring to it all these same concerns. Our battles in Placitas were predicated on the rights of the indigenous people who are most directly affected by growth and development that also threatens the natural world. While we may have occasionally stepped on each others' toes in trying to fight the good fight in an inclusive way, many of the alliances we formed have carried over to our work in the north."

What *has* changed—or intensified—are some perceptions: that there can be both wilderness and inhabited wilderness; that the land grants must achieve restitution to restore community management of their resources; and that commitment to the land is a necessary fight against globalization. My neighbors and the people I work with on a daily basis are the ones who have taught me this and have helped me understand why Sam Hitt and

I have come to have such a different perspective, which I question in my chapter *La Jicarita*. Maybe I now know why he cannot reconcile ecological/cultural landscapes and politics: he is physically separated from the cultural landscape—he doesn't go on roundups with his neighbors or clean his *acequia* every spring—and because his conception of an ecological landscape is biocentric. I believe this view of nature without people—"environmentalism-sin-gente"—results in a lack of political analysis that distorts the race, class, and gender lessons I learned in the 60s.

In Santa Fe, he surrounds himself with other white, middle-class environmentalists who support his ideas about conservation biology, which they believe provide all the solutions to our environmental problems. At a 1999 meeting sponsored by Rio Arriba County to discuss the content of Forest Guardians' ambitious Wildlands' recovery project, or the *State of the Southern Rockies: San Juan-Sangre de Cristo Bioregion*, Lorenzo Valdez, County Manager, asked that the group be more open to historically-proven cultural science. This cultural science has, after all, kept the northern New Mexico landscape relatively intact for 300 years. People on the land means having a "wisdom of the land," knowing what activity is appropriate for every part of your land: the wetlands next to the river for wildlife habitat; above that, *las vegas*, for summer pasture; then *la joya*, where the most fertile soil grows vegetables; and finally, *el altito*, just below the *acequia* where the fruit trees grow.

Hitt dismisses this kind of traditional knowledge as irrelevant in today's world. In his book, *Understories: The Political Life of Forests in Northern New Mexico*, Jake Kosek, now a professor of political geography at Berkeley, interviewed Hitt, who had this to say about *norteños*: "[T]hese people [Hispanos] are not traditional resource users but loggers and forest users like anyone else.... They may have once been traditional, but they've lost that now.... [T]he people's culture has been so contaminated by the dominant culture that they've lost any traditional ties to the land." He went on to say that "These forests belong to the whole country; they are not meant to serve as welfare for the people of Northern New Mexico."[93]

Unfortunately, as I look back on these highly charged politics of the 1990s from the vantage point of the 2000s, I see the cumulative damage environmental and forest policies have caused. La Compañía Ocho has ceased to exist, unable to meet its loan requirements. La Montaña de Truchas struggles to get even small restoration contracts from the Forest Service that barely keep the company viable. Max Córdova lost his position as president of the Truchas Land Grand in a bitter internecine fight over control of the grant. Efforts to organize a statewide community based forestry alliance floundered time and time again, as small, local loggers compete for federal monies earmarked for restoration work. Very little forest land is available for contract, as the Forest Service seems incapable of completing the necessary NEPA clearances in a timely fashion.

The Agua/Caballos timber sale in the Vallecitos Sustained Yield Unit lingered for over 12 years, tied up in appeals and lawsuits (by Forest Guardians, Carson Forest Watch, and Wild Watershed, a group founded by Sam Hitt after he left Forest Guardians), rewrites, and numerous reductions in board feet offered. Las Comunidades, one of the few designated operators extant in the Unit, lacks the capacity to bid on any kind of sawtimber contract. According to Ike DeVargas, after Las Comunidades received the Vallecitos sawmill, it rejected all offers from La Companía to lease the mill or contract with them to saw the timber from La Companía's sales. In an article in *La Jicarita News*, DeVargas, who is now retired due to the demise of La Companía and multiple health problems, had this to say about what has happened in the Unit:

"The Forest Service and the environmentalists have succeeded in getting the people of the Sustained Yield Unit villages fighting each other. And until we find a unity of purpose, we will have *nada*. We're a community divided, beating each other up for the crumbs."[94]

We all have things to learn as we try to find our way against forces hell-bent on marginalizing us even further. The damage wreaked by the timber barons and mining conglomerates of the 20th century pales in comparison to what may be in store for us as global forces continue to abrogate not only the sovereignty of indigenous cultures but county and state jurisdictions as well. The mantra "think globally act locally" seems more an oxymoron than a possible code of conduct.

Until all of us who care about the land in northern New Mexico cease fighting each other we will never find the *fuerte*, the strength, to survive. I tried to say it in an essay I wrote for *The Santa Fe Reporter* called "Who Would Have Thunk?" I talked about some of the strange alliances that have been formed over the years as a result of the fallout between *norteño* activists and certain environmental groups. This is how I ended the piece: "Many of us are ready to leave the bickering and intransigence behind, to work towards maintaining the integrity of rural communities and the wilderness they inhabit.... [W]e need to continue the discussion of how certain laws are affecting the socioeconomic needs of land-based people without gutting the laws themselves, which provide very real and valuable environmental protections. The only way to do this is to sit down with everyone necessary to resolve these issues—you don't really have to go to bed with them—and let go of the self-righteousness that comes from assuming you have all the answers. After having lived all of my adult life in this unique, diverse, and incredibly complex community of northern New Mexico, I have realized that I don't. Who would have thunk?"

Author on the roof of her house in Placitas. Photograph by Celia Jordan.

Notes

1. Conversation with Lizzie Archibecque, Placitas, NM, 1985.
2. Conversation with John Nordstrom, Placitas, NM, 1985.
3. Conversation with Lizzie Archibecque.
4. Larry Goodell, "Take Me For a Walk in Placitas," 1986.
5. Dan Flores, "The West that was, and the West that can be," *High Country News*, August 18, 1997, pp. 1, 6-7.
6. Kay Matthews, "Land Use & Lifestyle," *Albuquerque Journal Impact Magazine*, February 24, 1987, p. 10.
7. Ibid., p. 11.
8. Author(s) unknown, *Placitas Unreal Estate News*, 1987.
9. Lynn Montgomery, letter to Placitas Land Company, Placitas, NM, September 1987.
10. Patricia O'Conner, "The Plactias Good Life," *Albuquerque Living*, July 1988, pp. 37-44.
11. Author(s) unknown, *Placitas: The High Life*, 1988.
12. Kay Matthews, "Land Use & Lifestyle," *Albuquerque Journal Impact Magazine*, February 24, 1987, p. 11.
13. Ibid., p. 11.
14. Ibid., p. 11.
15. Ibid., p.11.
16. Ibid., p. 12.
17. Ibid., p. 11.
18. Ibid., p. 12.
19. Ibid., p. 12.
20. Ibid., p. 12.
21. Ibid., p. 12.
22. Slim Randles, "Developers' Signs Extol Placitas Life," *Albuquerque Journel Metro Plus*, October 19, 1989, p. 3.
23. Kay Matthews, "Indian Land Claims Deserve Support," *High Country News*, September 21, 1992, p. 15.
24. Ibid., p. 15.
25. Ibid., p. 15.
26. Ibid., p. 15.

27. Ibid., p. 15.
28. Ibid., p. 15.
29. Slim Randles, "Developers' Signs Extol Plactias Life," *Albuquerque Journal Metro Plus,* October 19, 1989, p. 3.
30. Maureen Hightower, et al., "Citizens for Rural Placitas," November 14, 1988.
31. Ed Quillen, "Reporter's Notebook," *High Country News,* November 4, 1991, p. 16.
32. Donald Dale Jackson, "Around Los Ojos, sheep and land are fighting words," *Smithsonian,* April 1991, p. 43.
33. Ibid., p. 43.
34. Ibid., p. 46.
35. Ibid,. p. 46.
36. Interview with Luis Torres, Española, NM, 1996.
37. Ibid.
38. Interview with Joanie Berde, Llano San Juan, NM, January, 1996.
39. Ibid.
40. Ibid.
41. Interview with Luis Torres, Española, NM, 1996.
42. Interview with JoAnn Medina, Taos, NM, 1996.
43. Interviews with Kat Duff, Taos, NM, 1996.
44. Doug McClellan, "Ranchers lynch activists in effigy," *Albuquerque Journal North,* November 25, 1995, p. 1.
45. William deBuys, "Forest fight isn't 'us, 'them'—just us," *Santa Fe New Mexican,* December 21, 1995, p. A-7.
46. George Johnson, "In New Mexico, An Order on Elusive Owl Leaves Residents Angry," *New York Times,* November 26, p. 1. 16.
47. Kay Matthews, "A Meeting of Minds at the Oñate Center," *La Jicarita News,* September 1996, p. 1.
48. Ibid. p. 1.
49. Ibid. p. 4.
50. Ibid. p. 4.
51. Ibid., p. 4.
52. Ibid., p. 5.
53. Ibid., p. 5.
54. Max Córdova, Letter to Lori Osterstock, Española District Ranger, October 29, 1996.
55. Chellis Glendinning, et. al., "Inhabited Wilderenss," *Santa Fe New Mexican,* February 2, 1997.
56. Keith Easthouse, "Environmentalists toss Hitt out of group," *Santa Fe New Mexican,* April 8, 1997, Section B.

57. Julia Goldberg, "Environmentalists Lose Battle Over La Manga Sale," *Rio Grande Sun*, September 25, 1997, p. 1.

58. Field trip to La Manga timber sale, Vallecitos, NM, September, 1997.

59. Kay Matthews and Mark Schiler, "Interview with Antonio DeVargas," *La Jicarita News*, January 1996, pp. 6-8.

60. Kay Matthews and Mark Schiller, "Interview with Joanie Berde of Carson Forest Watch," *La Jicarita News*, February 1996, pp. 4-7.

61. Conversation with Praxedis Ortega, Vallecitos, NM, August 3, 1996.

62. Conversation with Sam Hitt, Vallecitos, NM, August 3, 1996.

63. Phone conversation with Ike DeVargas, August 1996.

64. Phone conversation with Sam Hitt, January 1997.

65. Steve Chase, editor, "Introduction," *Defending the Earth*, p. 20.

66. Sam Hitt, "Letters," *La Jicarita News*, March 1997, p. 3.

67. Courtney White, *Quivira Coalition Newsletter*, June 1997.

68. William deBuys and Alex Harris, "Water Will Show," *River of Traps*, p. 27.

69. Phone conversation with Bruce Hamilton, October 1997.

70. Courtney White, "The Uneasy Chair," *The Sierran*, May/June 1997, p.3.

71. David Orr, e-mail to Sierra club list serve, September 1997.

72. Courtney White, "Sierra Club Chapter Refuses to Participate in Bingaman's Roundtables," *La Jicarita News*, August 1998, p. 4.

73. Kay Matthews, "Sierra Club Hears From Minorities Locally and Nationally," *La Jicarita News*, August 1999, p. 7.

74. Ibid., p. 7.

75. Kay Matthews, "Norteños Claim National Policies of Environmental Groups Discriminate Against Minorities and Fail to Protect Resources," *La Jicarita News*, October 1999, pp. 1, 7.

76. Kay Matthews, " 'Zero Cut' Logging Initiative Results in Internal Conflict Among Environmental Organizations," *La Jicarita News*, November 1997, p. 5.

77. Chad Hansen e-mail to Earth Island Institute, "Environmental Justice?", October 10, 1997.

78. "Fight the Corporate Destruction of Public Lands, Not Land-Based Communities," Urban Habitat Program, November 11, 1997.

79. Kay Matthews, " 'Zero Cut' Logging Initiative Results in Internal Conflict Among Environmental Organizations," *La Jicarita News*, November 1997, p. 5.

80. Ralph Schmidt, "Letters to the editor," *New York Times*, August 6, 1997.

81. Kay Matthews, "Sierra Club Votes on Restricting Immigration to Control Population," *La Jicarita News*, May 1998, p. 4.

82. Ibid., p. 4.

83. Ibid., pp. 4-5.

84. Ibid., p. 4.

85. George Sessions, "The Sierra Club, Immigration & the Future of California," *Wild Duck Review*, Winter 1998, pp. 24-25.

86. Ibid., p. 5.

87. Dave Foreman, "Progressive Cornucopia," *Wild Duck Review*, Winter 1998, pp. 16-17.

88. Karen Smith e-mail to *La Jicarita News*, November 9, 1997.

89. Courtney White, "Sierra Club Chapter Refuses to Participate in Bingaman's Roundtables," *La Jicarita News*, August 1998, p. 4

90. Ernie Atencio, *Of Land and Culture: Environmental Justice and Public Lands Ranching in Northern New Mexico*, A Report by the Quivira Coalition and the Santa Fe Group of the Sierra Club, January 2001.

91. John Horning e-mail to Kay Matthews, October 31, 1997.

92. Mark Schiller, "Peñasco Grazing Workshop: Establishing Trust," *La Jicarita News*, July 1998, p. 1.

93. Jake Kosek, *Understories: The Political Life of Forests in Northern New Mexico*, Durham: Duke University Press, 2006.

94. Kay Matthews and Mark Schiller, "Editorial: Community Forestry, Troubled Times in Northern New Mexico," *La Jicarita News*, November 2004, p. 7.

www.ingramcontent.com/pod-product-compliance
Lightning Source LLC
Chambersburg PA
CBHW020532270326
41927CB00006B/543